BEER, BABES, and BALLS

SUNY series on Sport, Culture, and Social Relations

CL Cole and Michael A. Messner, editors

BEER, BABES, and BALLS

Masculinity and Sports Talk Radio

David Nylund

Foreword by Eric Anderson

STATE UNIVERSITY OF NEW YORK PRESS

Some material in chapters 5 and 6 was originally published in David Nylund, *The Journal of Sport and Social Issues* 28(2), pp. 136–140 and 166. Copyright ©2004 by Sage Publications. Reprinted by permission of Sage Publications.

Published by
State University of New York Press, Albany

© 2007 State University of New York

All rights reserved

Printed in the United States of America

No part of this book may be used or reproduced in any manner whatsoever without written permission. No part of this book may be stored in a retrieval system or transmitted in any form or by any means including electronic, electrostatic, magnetic tape, mechanical, photocopying, recording, or otherwise without the prior permission in writing of the publisher.

For information, contact State University of New York Press, Albany, NY
www.sunypress.edu

Production by Marilyn P. Semerad
Marketing by Michael Campochiaro

Library of Congress Cataloging-in-Publication Data

Nylund, David.
  Beer, babes, and balls : masculinity and sports talk radio / David
Nylund ; foreword by Eric Anderson.
    p.   cm. — (Suny series on sport, culture, and social relations)
  Includes bibliographical references and index.
  ISBN 978-0-7914-7237-8 (hardcover : alk. paper) —
  ISBN 978-0-7914-7238-5 (pbk. : alk. paper)
  1. Radio broadcasting of sports—United States.
  2. Radio talk shows—United States. 3. Radio and baseball.
  4. Men—United States—Attitudes. 5. Masculinity—United States. I. Title.
  GV742.3.N95 2007
  070.4'497960973—dc22   2006037455

10 9 8 7 6 5 4 3 2 1

*To Debora Bubb for all her incredible sacrifice and love
while I focused on this book*

CONTENTS

The author, Dr. David Nylund, could have asked a big-name sports personality or on-air radio sportscaster to write the foreword to this book. It is, after all, a work written mostly about white, heterosexual men who love sports. So it is significant that Dr. Nylund has asked me, an openly gay academic, extremely critical of the way sports marginalize gay men, women, and often people of color. Whereas Dr. Nylund reveres sports talk radio so much that he is likely to seek traffic jams to increase his listening time, I egotistically confess that I only listen if I am being interviewed on the show. Sports, to me, are the opiate of the masses. I hate sports so thoroughly that I turn the dial the moment I hear the words "sports are next." Yet, after reading *Beers, Babes and Balls*, I admit to having sought out a few of those AM stations, largely to see if the author's findings resonate with my own.

Just over a year ago I was invited to speak on dozens of sports talk shows about my book, *In the Game: Gay Athletes and the Cult of Masculinity*. One morning I was talking to Deter, an on-air radio personality from a city far from the liberal comfort of Los Angeles or England, where I split my time. Deter resembles the typical sports fan. He has no clothing style, and, other than an emerging belly, he looks like he once ran with or threw a ball for some team. Deter tells me that he is really intrigued by my research on openly gay athletes. Like most of the sportscasters I talked with, he was clear to preface that he had no problem with gay men—or gay athletes, by extension. But of all

Dr. Eric Anderson came out of the closet as America's first openly gay high school coach in 1993. Today he teaches in the Department of Education at the University of Bath, England. He has written an autobiography, *Trailblazing: The True Story of America's First Openly Gay High School Coach*, and is the author of a book about openly gay and closeted gay athletes in team sports, *In the Game: Gay Athletes and the Cult of Masculinity*. You can find out more about him and his research at EricAndersonPhD.com.

the sportscasters who interviewed me, Deter stands out because he tells me that he has several gay friends and is a huge supporter of gay rights. After a few minutes of off-air liberal conversation, Deter notes "My on-air personality is quite different, though."

Deter informs me that he is known for his brash, brazen, and (much to my surprise) homophobic sentiments. "It's an act," he says. And, just a minute before we go on air, he asks if I'm willing to "roll with it." With no real opportunity to say no, I reply, "I warn you, I will give it back." Deter smiles, the appropriate machines beep and flash, and Deter's voice suddenly morphs into a hypermacho Sunday, Sunday, Sunday type of voice: "We're talking with Eric Anderson, author of a book about openly gay athletes." As so many do when the topic of gay men comes up, Deter is quick to sexualize my sexuality. "The thought of two men doing that just makes me sick," he says; "the thought of it makes me want to throw up." "Go ahead," I say. "Why do homophobic men say they want to throw up, but never do? If it really grosses you out that much," I chide him, "put your fingers where your mouth is and throw up!" Much to my surprise, Deter reaches his two fingers back deep into his throat and makes a few gagging sounds. He removes his fingers, saliva stretched between them. "That's the deepest I've ever seen a straight man put something down his throat without gagging," I say. "You're sure you are straight?" Everyone in the studio laughs, and Deter goes to commercial.

I received an email the next week from Deter saying that he has been replaying that bit every day for a week and that it's really helped his ratings. While I am, at one level, glad to help Deter, I am also struck by an awful thought: is this what we (gay men) have become within the sports media complex—a punching bag for those in the conservative entertainment business to increase their ratings, even if they do not believe in what they espouse? In person, Deter tells me that he is a huge supporter of gay rights, but, on the air, he utilizes his agency to influence the masses to hold homophobic sentiments.

Deter's transformation from gay rights supporter to raging homophobe is reminiscent of many heterosexual team sport athletes I work with. When speaking to these athletes privately, few (very few) ever say that they are homophobic, and most insist they would have no problems with a gay teammate. However, they all maintain that their teammates would. The result: team sport athletes often publicly declare homophobia, even though they disagree with it privately. Whereas I research this question among current athletes, Dr. Nylund has extended this investigation into an arena for those no longer play-ing sports: sports talk radio.

I have previously described team sports as a total institution. What I mean by this is that, from the time boys enter sports, their social worlds begin to close in around just sports. The longer they remain in sports, the more their social networks, time, and personal identities are consumed and shaped by sports. By the time an athlete attends college, he is generally living with, practicing with, traveling with, tak-ing classes with, and partying with his teammates. His social world is closed to other people's narratives and he identifies as a jock. But what about when a jock's career ends? What happens to the millions of team sport athletes who, through injury, de-selection, or graduation end up having to disassociate with team sports and work a 9 to 5 job?

Beer, Babes, and Balls is an exploration into just this question. Working from the premise that sports radio is a "third place" for white, heterosexual men, Nylund shows how these men reconnect with their former identities as jocks. He demonstrates how sports radio serves as a community, real for some, imagined for others, of men attempting to relocate themselves with the privileged status of white, heterosexual jock as well as a place in which those who always wanted to be a jock, but did not have the talent, can air their voice on an equal playing field with those who did play. Sports radio and sports bars therefore become places where white, heterosexual men bond over the objectification of women, the subordination of gay men, and the occasional marginalization of athletes of color or those from lower socioeconomic class. Thus, whereas gay men, women, and peo-ple of color are increasingly making gains in once patriarchal, homog-enous, and homophobic institutions like the church, government, mil-itary, and workplace, in sports talk radio (and with the help of a call screener), men can again be men, and all one needs to do to belong to this community is bond in the opposition to homosexuality and femi-ninity by espousing the virtue of beer, babes, and balls.

But, as Nylund aptly points out, men reeling in their disassociation from the patriarchy their fathers once promised them are not totally safe from the gains of women and gay men, not even in sports radio; there are cracks and fissures even in this conservative world. Largely focusing on one show, *The Jim Rome Show*, Nylund points out that, by conservative standards, Rome is somewhat progressive. While I might not lump Rome into the net of liberal commentators, Nylund points to Rome's (often conflicting) broadcasts that, importantly, his callers perceive as being pro-gay. This begs the question. Is Rome's "less homophobic and less misogynistic" tone the right tenor for the times? Is it possible that Rome presents the right combination of conservatism in order to—slowly—bring his audience into a more enlightened place? Perhaps Rome is trying to make slow, progressive social change from inside the belly of the beast.

But there is another story to this book. It is equally as fascinating a read of how a sport scholar, recently trained in the critical analysis of sport, rectifies his lifelong love affair with his addiction to sports talk and sports radio. Academics will thus find that his book serves as a powerful case study for how sport scholars (who are almost exclusively ex-athletes themselves) deal with the interpretation and meaning of their findings and the influence that their own bias brings to their work. For nonacademics, Nylund's self-reflexivity (largely because he is also trained as a psychotherapist) of both loving sport and sports radio, but standing firmly for his pro-gay, pro-women, and pro-racial equality beliefs, might position him as the perfect spokesperson for opening up the airwaves, the third space, for the inclusion of those previously marginalized by sport.

ACKNOWLEDGMENTS

There are so many people to thank—friends, family, mentors, peers, and faculty who have supported me financially, academically, spiritually, and emotionally—and too many to mention them all by name. So, if I forgot to thank you, please know that I sincerely appreciate all you have done. There are some people, however, that I particularly want to thank.

First, I am grateful to my professors at UC Davis: Judith Newton, Laura Grindstaff, Jay Mechling, Susan Kaiser, Gayatri Gopinath, Ruth Frankenburg, Catherine Kudlick, Dean MacCannell, Kent Ono, Anna Kuhn, Sarah Projansky, and Karen Shimakawa—your ideas have had a profound influence on me. Thanks to my colleagues at California State University, Sacramento, who helped sustain me this year while I wrote this book: Marilyn Hopkins, Pat Clark-Ellis, Robin Carter, Wandarah Anderson, Joe Anderson, Mimi Lewis, Tony Platt, David Demetral, Sylvia Navari, Susan Eggman, Robin Kennedy, Robin Carter, Susan Taylor, Jill Kelly, Maria Dinis, Chrys Barranti, Dale Russell, Janice Gagerman, Lynn Cooper, Francis Yuen, and Andy Bein.

A special thanks to friends who supported me during the writing of this book: Jeff Chang, Lisa Otero, Jesse Patrone, Michael Davis, Craig Smith, Stephen Madigan, Colin Sanders, Mauricio Vargas, Diana Thant, Avery Malby, Roxana Borrego, Vic Corsiglia, Rosemary Hill, Sarah Bateni, Antonia Taylor, Rosemary Madruga, Kenny Benjamin, Reed Walker, Leah Walls, David Elliott, Cesar Castenada, Poshi Mikalson, and Jon Kosier. Special thanks to my dear friends Julie Tilsen and Lauri Applebaum for allowing me to stay with them in Minneapolis to finish the book. A very special thanks to my son Drake for his patience. And to Debora, for all your sacrifices so I could focus on this book; I am eternally grateful. Also, I am indebted to SUNY Press, particularly Nancy Ellegate, for believing in this project. I offer thanks for coming through in the clutch on last-minute notice. I am also grateful

xvi   *Beer, Babes, and Balls*

to Michael Messner and CL Cole, editors of the SUNY Press Sport, Culture, and Social Relations Series, for their support.

I am particularly appreciative of my favorite bartenders in the sports bar world: Heather Evangelista and Angie Soares. Finally, I want to thank all the fans of sports radio who agreed to be interviewed. I learned a great deal and enjoyed my time hanging out with you all. Thanks to sports 1140 AM, Sacramento, for your support of my research. I also want to give a "shout out" to all the sports bars I frequented. It was a health sacrifice—I gained twenty pounds while conducting this research due to all the bar fare I consumed (and props to Jon Kosier for helping me lose the weight!). I have to go on a diet— from sports bar food, not sports talk radio.

1 1st Inning

OPENING PITCH

Thinking about Sports Talk Radio

I am driving in traffic on a typical harried Monday morning. Turned
off by the conservative "hate speech" of political talk radio and bored
by Bob Edwards of NPR, I turn on my local sports radio station. A
commercial plugging the local station is airing: "Your hair's getting
thinner, your paunch is getting bigger. But you still think the young
babes want you! That's because you listen to Sports 1140 AM—it's
not just sports talk, it's culture." Next comes the loud, rhythmic guitar
riff from a Guns N' Roses song, "Welcome to the Jungle." As Axel
Rose belts out the lyrics, an announcer bellows, "Live from Los Angeles, you're listening to *The Jim Rome Show.*" Next, the distinct,
brash voice of Jim Rome, the nation's most popular sports talk radio
host, addresses his audience of two million sports fans[1]: "Welcome
back to the Jungle. I am Van Smack. We have open phone lines. But
clones, if you call, have a take and do not suck or you will get run."[2]
Over the next three hours, the well-known host interviews famous
sports figures, articulates controversial opinions, and takes phone
calls from his loyal listeners/sports fans who speak in such Rome-
invented terms as "jungle dweller," "bang" and "Bugeater."[3] I listen to
the program with mixed feelings. As a sports fan, I find myself en-
grossed and amused; I want to know what each "in-group" term
means. As a critical feminist scholar, I am uneasy with his confronta-
tional and insulting style, not to mention the aggressive and uncritical
content of his speech. I wonder, "What will Rome say next?"[4]

The Jim Rome Show reflects a growing cultural trend in the United
States—sports talk radio. The most popular sports talk radio hosts, including Jim Rome and others such as Mad Dog and JT the Brick, built

their reputations through their obnoxious and combative styles. With white male masculinity being challenged by feminism, affirmative action, gay and lesbian movements, and other groups' quests for social equality, sports talk shows have become an attractive venue for embattled white men seeking recreational repose and a nostalgic return to a prefeminist ideal (Farred, 2000).

SPORTS TALK RADIO

Presented as a medium in which citizens/callers can freely "air their point of view," talk radio has become a popular forum for large numbers of people to engage in debate about politics, religion, and sports. The media culture, with talk radio as a prominent discourse, plays a very powerful role in the constitution of everyday life, shaping our political values and gender ideologies, and supplying the material out of which people fashion their identities (Kellner, 1995). (Hence, it is crucial for scholars to furnish critical commentary on talk radio; specifically, we should critique those radio texts that work to reinforce inequality.)

Talk radio formats, particularly political talk radio, exploded in the 1980s as a result of deregulation, corporatization of radio, and niche marketing (Cook, 2001).[5] Deregulation, which loosened mass-media ownership and content restrictions, both renewed interest in radio as a capitalist investment and galvanized the eventual emergence of its two 1990s prominent showcase formats: "hate" talk radio shows and all sports programming (Cook, 2001). By the late 1990s, there were more than 4,000 talk shows on 1,200 stations (Goldberg, 1998).[6] Sports talk radio formats have, according to cultural studies scholar Jorge Mariscal (1999), "spread like an unchecked virus" (p. 111). Currently, there are over 250 all-sports stations in the United States (Ghosh, 1999).

As a result of deregulation and global capitalism, new media conglomerates emerged as the only qualified buyers of radio programming. Infinity Broadcasting, the largest U.S. company devoted exclusively to owning and operating radio stations, owns WFAN and many other sports radio stations. Its competitor, Premiere Radio Network, owns the popular nationally syndicated programs formerly hosted by Howard Stern (who went to satellite radio—Sirius), and currently hosted by Rush Limbaugh, Dr. Laura, and Jim Rome. Herbert Schiller (1989) refers to this programming trend as "corporate speech" (p. 40) that encourages censorship and contains public expression within corporate, capitalist ideologies that reinforce dominant social institutions.

With the growing corporate ownership of radio came niche marketing that caters to targeted demographic groups. Talk radio is aimed at a specific demographic: white middle-class men between the ages of twenty-four and fifty-five. Research shows that talk radio listeners are overwhelmingly men who tend to vote Republican (Hutchby, 1996; Page and Tannenbaum, 1996). The most popular program, *The Rush Limbaugh Show*, has twenty million daily listeners who laugh along with the host as he rants and vents, opening a channel for the performance of the "angry white male." Roediger (1996) remarks, in a fascinating reading of Limbaugh's cultural significance in the United States, that "banality can carry much more social power than genius where White consciousness is concerned" (p. 42). Susan Douglas (2002) argues that while most research on talk radio focuses on the threat it poses to democracy, what is obvious but far less discussed is talk radio's central role in restoring masculine hegemony. Similarly, sports talk radio, according to Goldberg (1998), enacts its white dominance via hypermasculine posing, forceful opinions, and loud-mouthed shouting. Sports talk radio "pontificates, moralizes, politicizes, commercializes, and commodifies—as it entertains" (Goldberg, p. 213). Although Rome's masculine style is different from Limbaugh's and Stern's, all three controversial hosts have built reputations through their rambunctious, masculinist, and combative styles.

Beers, Babes, and Balls explores the burgeoning genre of sports talk radio and related ideas about masculinity. It provides an interdisciplinary approach to this subject, drawing from sports sociology, media and cultural studies, masculinity studies, and queer studies. Popular culture plays a significant role in the fashioning of (post)modern identities, and sports talk radio is both a representative site and an organizing force of important cultural shifts in masculinity. The focus on sports radio sheds light on certain aspects of contemporary masculinity and recent shifts in gender and sexual politics. This book also examines sports talk radio within a broader cultural context that includes television, film, and men's magazines.

In writing this book, I wanted to answer the following questions:

- How is masculinity and its intersection with race, class, gender, and nation represented in sport talk radio? What do the images of masculinity portrayed in sports radio reveal about current gender politics?

- What accounts for sports talk radio's appeal? What kind of bonding occurs in the mediated space of sports talk radio? Who is included in this airwave community and who is excluded?
- What is the link between portrayals of masculinity and ideas about heterosexuality in sports talk?
- How can we best articulate the relationship between the production, text, and consumption of sports talk radio?
- Where does sports talk radio fit within the larger context of popular culture and the relationship between popular culture and consumerism?

Why study this topic and what significance does this topic have for me? I became attracted to this subject for two reasons. First, as a cultural studies scholar, I am interested in studying cultural practices of everyday living (such as listening to sports talk radio). Second, as a sports fan who has been a regular listener to sports talk radio, I am curious to examine a genre in which I participate. Because I write both as a scholar and as a fan, this book reflects these two levels of knowledge, which are not necessarily in conflict but also are not necessarily in perfect alliance. Being a fan allows me certain insights into sports talk radio that an academic who is not a fan might not have, particularly when that analysis of texts is isolated from actual audiences. I thus avoid Jenkins's (1992) critique of academic textual analysis that is distant from audiences and consequently "unable to link ideological criticism with an acknowledgment of the pleasures we find within popular texts" (p. 7). Because I am a fan, I am participating in the subject of my study, which has implications (as well as constraints) for what I observe and understand about the topic. Next, I will discuss the conceptual frameworks that inform this book, keeping in mind the benefits and perils of fandom.

Just as I am both sports fan and scholar, this book is written for both sports fans and academics. Warning, sports fans: you may want to skip some of the theoretical discussions in this book. Theory is important for us professor types, but may not be so interesting or necessary for general readers to benefit from and understand this book. There is enough material here that you will comprehend and connect with as a sports fan. However, I also caution general readers and sport talk radio listeners: as a cultural studies and feminist researcher, I am committed to promoting a just and democratic society. This commitment means social justice for women, people of color, and gays and lesbians. This book may confront some of your taken-for-granted ideas about sports, particularly the idea that sports exists outside politics and power. In addition, this book may challenge male sports fans to examine their ideas of masculinity and invite them into accountability about male dominance, homophobia, racism, and sexism. Enough said. I will now discuss the conceptual frameworks that inform this book, keeping in mind the benefits and perils of fandom.

THEORIZING MASCULINITIES

I view masculinity as a social construction that assumes different forms in different historical moments and contexts (Jackson, Stevenson, and Brooks, 2001). Sociologists have long recognized that there are diverse forms of masculinities found among different cultures (Anderson, 2005). What it means to be masculine shifts also within the same culture over time due to various political, social, and economic forces, and not all masculinities are treated similarly. Connell, in his book, *Masculinities,* describes the various and often conflicting forms of masculinities in Western societies, particularly for understanding the operation of hegemony as it relates to masculinity.

Hegemony, a theory developed by social theorist and activist Antonio Gramsci (1971), refers to a form of dominance in which the ruling class legitimates its support from and power over the subordinate classes not by brutal force but through more insidious forms of control and consent (e.g., through the media, institutions, and schools). The subordinated classes, through the process of hegemony, come to see their marginal places as both right and natural. Eric Anderson (2005) states that hegemony is exemplified when

> a slave believes his rightful place is that of a slave (a racist society), when a woman believes she should be subservient to a man (a sexist society), when a poor person believes that he does not merit wealth (a classist society), or when a gay man believes he is undeserving of the same rights as a straight man (a heterosexist society). (p. 21)

Anderson goes on to note that hegemony has been a central concept in understanding oppression of racial minorities, women, gays and lesbians, and the lower classes in Western society. Recently, the concept of hegemony has been applied to a more complex understanding of how men and their masculinity are stratified in society (Anderson, 2005).

Much of masculinities studies centers on how men construct hierarchies that yield decreasing benefits the farther removed one is from the ideal version, something identified as hegemonic masculinity. Gender scholars have described at least five distinctive features of hegemonic masculinity in U.S. culture: (1) physical force, (2) occupational achievement, (3) patriarchy, (4) frontiermanship, and (5) heterosexuality (Brod, 1987; Kimmel, 1994). Connell (1990) defines hegemonic masculinity as "the culturally idealized form of masculine character" (p. 83) that emphasizes "the connecting of masculinity to toughness and competitiveness," as well as "the subordination of women" and "marginalization of gay men" (p. 94). Connell also suggests that hegemonic masculinity is not a static phenomenon, but an always-contested, historically situated social practice.

Michael Messner (1997) has a useful framework for theorizing masculinities in a U.S. context:

1. Men, as a group, enjoy institutional privileges at the expense of women, as a group.
2. Men share very unequally in the fruits of male privilege/patriarchy: normative/hegemonic masculinity (white, middle- and upper-class, heterosexual) is constructed in relation to femininities and to various subordinated masculinities (racial, sexual, class, female masculinity).
3. Men can pay a cost—in the form of poor health, shallow/narrow relationships, for instance—for conformity with the narrow definitions of masculinity that promise to bring them status and privilege.

Messner's thematics allow theorists to speak of masculinities in the plural and to put the relationship between gender and power at the center of analysis. Furthermore, his conceptualization creates space to examine connections between the construction of masculinities and other social constructions, such as race, class, and sexuality.

In addition to studying the construction of gender (accessed through my textual analysis and audience interviewing), I am influenced by ethnographic approaches to masculinity (Duneier, 1992; Cornwall and Lindisfarne, 1994). Hence, this book will move beyond research on masculinity and the media (Craig, 1992) in which the level of analysis remains at the level of the textual and grapple with how these media representations are negotiated by individual and groups of men (in my case, sports talk radio fans).

Ethnographic approaches to masculinity are needed to understand men's actual lived experience of masculinities. Without researching masculinities in real-life settings, including sport, the research is reduced to merely theoretical understandings of manhood. Surely, masculinity studies has been invaluable in unmasking an historically invisible category, particularly through writings from people who have been marginalized by white male dominance—women, gay men, people of color. I am indebted to such writings (e.g., hooks, 1984). Yet what jumps out in much of the writing on masculinity is the overall negativity and its bleak outlook regarding men. Beynon (2002) comments on this overemphasis on pessimism in masculinities scholarship and writings: "A Martian arriving on Planet Earth and not knowing what masculinity was would quickly form the opinion that it is a highly damaged and damaging condition with very few, if any, redeeming features" (p. 143). This book addresses this problem by exploring how real men experience contemporary masculinity, how they perform masculinity, how they connect to other men and women, and how men rework ideals of masculinity in a postmodern society. Thus, I hope to examine how the proliferation of masculinities may open up new opportunities for men.

In addition to being positioned as a cultural studies informed ethnographer, I am situated as a psychotherapist who has worked in various clinical settings for the past two decades. Most of my research and writings has been in the area of narrative therapy (Nylund, 2000; White and Epston, 1990), a school of therapy that is informed by poststructuralism, critical theory, and feminist theory. The goal of narrative therapy is to deconstruct the client's problem-saturated story and support the client in noticing and performing alternative and preferred stories. White and Epston (1990) suggest that no single story accounts for the person's total lived experience since there are always contradictory stories and accounts of a person's life. These alternative stories are referred to in narrative therapy as "unique outcomes"— events that contradict or resist the dominant problem narrative. In a clinical interview, unique outcomes serve as an entry point into alternative stories and pathways into new ways of being. A narrative therapy framework is useful in my analysis of sports radio and masculinity. I intend to be vigilant to unique outcomes in sports radio—moments that contradict sports radio's dominant themes of misogyny, nationalism, racism, or homophobia. These moments, I will suggest, should not be minimized since they can serve as an entry point for substantively discussing sexism, racism, and heterosexism.

DEVELOPMENT OF MANHOOD IN
TWENTIETH-CENTURY UNITED STATES

In order to understand contemporary masculinity, it is imperative to give a historical context of the shifting nature of ideas of manhood. The late nineteenth-century American male image was that of a rugged individualist who, to escape civilizing constraints, went to work in exclusive male preserves, went to war with other men, and went West to find fortune, pitting his will against the perils of nature (Kimmel, 1996). However, as the United States became increasingly urban and mobile in the early twentieth century, these "masculine" options were no longer available, and men were forced to look elsewhere to reclaim their lost identities. To many middle-class white men, this retrieval of identity was vital due to the changing nature of work, the visibility of first-wave feminism, the closing of the frontier, and changes in family relations (e.g., modern urban boys being separated from their fathers and placed in the care of mothers or women schoolteachers). The resultant changes in work and family life brought on by urbanization led to fears of boys and men being feminized. Many men, in response to these changes, searched for places "where they could be real men with other men" (Kimmel, 1996, p. 309) and where they could actively exclude women, nonnative-born whites, men of color, and homosexuals. Men created homosocial organizations (male-only spaces) such as fraternal lodges, rodeos, college fraternities, and the Boy Scouts to initiate the next generation of traditional manhood (Mechling, 2001; Messner, 1997).

Beginning with the 1960s, a similar pattern of work and family relations, along with second-wave feminism, civil rights, and the gay liberation movement, has produced another so-called period of crisis or confusion for men (Kimmel, 1996). Much like their turn-of-the-century ancestors, men began renewing efforts to reinvent masculinity. These renewal efforts took the forms of various "men's movements": men's rights advocates, feminist men, men of color, gay male liberationists, Promise Keepers, and the mythopoetic men's movement (Messner, 1997). While some groups were viewed as "essentialist retreats" (Messner, 1997, p. 17) to restore a "true" manhood, all of these groups were attempting to make sense of masculinity in the shift from an industrial to information society and have been attempting to reinvent new ideals of masculinity (Newton, 2005).

Turning now to the late twentieth/early twenty-first century, masculinity once again was transformed, this time by commercial forces and neoliberalism (the weakening of the welfare state and consolidation of corporate power). The postmodern transformation of masculinity produced two new forms of commercialized masculinities: the "new man" and the "new lad" (Beynon, 2002; Nixon, 1997). Put briefly, the "new man" emerged in the 1980s around the time of contemporary U.K. and U.S. men's lifestyle magazines such as *Arena* and *GQ*, which produced a "new politics of looking as the 'male-on-male' gaze joined the 'male-on-female' as socially acceptable" (Beynon, 2002). Nixon has argued that the new politics of looking helped to challenge the previously unmarked or invisible status of men. The "new man" was an avid consumer and narcissist who also internalized and endorsed principles of feminism, including a reassessment of the traditional household division of labor and a new commitment to fatherhood (Beynon, 2002). The 1990s "new lad" was a clear reaction to the "new man" and arguably an attempt to reassert hegemonic masculinity deemed to have been lost by the concessions made to feminism by the "new man." "New laddism" is most clearly embodied in current men's magazines, such as *Maxim, FHM,* and *Loaded,* and marked by a return to hegemonic masculine values of sexism, male homosociality, and homophobia. Its key distinction from hegemonic masculinity was a huge dose of irony and reflexivity about its own condition that arguably rendered it immune from feminist criticism. Lastly, the "new lad" was also a construct that drew upon and appropriated working-class male culture for its values; was younger than the "new man"; was less invested in work, preferring to drink, party, and watch sports; made hardly any references to fatherhood; and addressed women as sexual objects.

I am interested in situating sports talk radio in the framework of the aforementioned historical manifestations of masculinity. It is my belief that, similar to developments in the early twentieth century, contemporary sports talk radio helps to construct a new male identity in response to changes in the gender and economic order. There are reasons for this. First, in the United States, sport is a key symbol for masculinity. Secondly, sports talk radio is a unique medium because the highly competitive nature of the radio market makes it likely that producers and program directors will exploit current cultural trends as quickly as possible. I am interested in exploring the ways that sports talk radio borrows from and exploits contemporary masculine ideals—namely, the versions of masculinity enacted by some of the men's movements and by the "new lad" and "new man." I am also interested in analyzing the type of male community that is created by sports talk radio: In what ways is it similar or dissimilar to other male homosocial communities?

MASCULINITY AND THE SPORTS MEDIA

Historically, sports have played a fundamental role in the construction and maintenance of hegemonic masculinity in the United States (Messner, 1992). Communications scholar Nick Trujillo (1996) states, "No other institution in American culture has influenced our sense of masculinity more than sport" (p. 183). The mass media have benefited from institutionalized sports and have served to reaffirm certain features of hegemonic masculinity. As Trujillo (1994) writes,

> Media coverage of sports reinforces traditional masculinity in at least three ways. It privileges the masculine over the feminine or homosexual image by linking it to a sense of positive cultural values. It depicts the masculine image as "natural" or conventional, while showing alternative images as unconventional or deviant. And it personalizes traditional masculinity by elevating its representatives to places of heroism and denigrating strong females or homosexuals. (p. 97)

Mediated sports texts function largely to reproduce the idea that traditional masculinity and heterosexuality are natural and universal rather than socially constructed (Jhally, 1989). Since these dominant texts have detrimental effects on women, gays, lesbians, and some men, Trujillo argues that mediated sport should be analyzed and critiqued.

Many scholars have taken up Trujillo's call and in the last decade we have seen an explosion of research on sports and mass media (Wenner, 1998a). Most of these studies examine televised sports and its link to violent masculinity, sexism, and homophobia (Messner, Dunbar, and Hunt, 2000). However, scholars have also turned their attention to the impact and meaning of "sports talk." Grant Farred (2000) describes sports talk as an "overwhelmingly masculinist (but not exclusively male), combative, passionate, and apparently open ended discourse" (p. 101). Farred characterizes sports radio talk shows as "orchestrated and mediated by rambunctious hosts" providing a "robust, opinionated, and sometimes humorous forum for talking about sport" (p. 116). Likewise, Sabo and Jansen (1998) posit that sports talk serves as an important primer for gender socialization in current times. They write:

> Sports talk, which today usually means talk about mediated sports, is one of the only remaining discursive spaces where men of all social classes and ethnic groups directly discuss such values as discipline, skill, courage, competition, loyalty, fairness, teamwork, hierarchy, and achievement. Sports and sports fandom are also sites of male bonding. (p. 205)

Sports radio does appear to have a communal function and is a particularly interesting site to study how men perform relationships and community. Pamela Haag (1996) finds something inherently democratizing about sports talk radio because she thinks it promotes civic discourse and teaches us how to create community "for a lot of people who lead isolated, often lonely lives in America" (p. 460). Haag also suggests that sports talk radio serves a function different from political talk radio, despite serving a similar, largely white middle-class audience, because the values that it emphasizes focus on community, loyalty, and decency. The appeal of sports talk radio, according to Haag, lies in the idiosyncrasies of its hosts and the regionalism of the issues covered, in direct opposition to the increased national corporate control of radio. Farred (2000), in speaking to the communal function of sports, suggests that sports talk on the radio can momentarily break down barriers of race, ethnicity, and class.

While acknowledging the productive potential of sports talk, Susan Douglas (2002) argues that talk radio (including sports talk radio) is to be understood as another attempt to retain certain aspects of hegemonic male identity that have been lost due to feminism. Its popularity with men coincides with other current media trends, including men's magazines such as *Maxim* and *FHM*, or Comedy Central Cable Network's hypermasculine television show, *The Man Show*, which, according to Maureen Smith (2002), represents a nostalgic (and perhaps ironic) attempt to return to a prefeminist masculine ideal. In particular, white, middle-class, heterosexual men may feel threatened and uncertain about changes encouraged by feminism and by gay rights. In a *Sacramento Bee* article titled "Frat Boy Nation: A New Culture of Chauvinism Buries the Sensitive Guy," popular culture writer J. Freedom du Lac (2002) identifies a new (or recycled) form of a masculinity being marketed to white men ages eighteen to forty. This mediated masculinity, similar to the "new lad," is typically seen as a backlash against the sensitive, pro-feminist male. It is characterized by a recycled, nostalgic form of masculinity, a throwback to a time when men were able to behave badly and not worry about censure. Men, according to this media-created construction, should return to a traditional, phallocentric masculinity that includes displays of sexism, consuming large quantities of alcohol, machismo, and hedonism—a return to a "frat boy nation." Television shows such as Fox Sport Net's *Best Damn Sports Show Period* and Comedy Central's *The Man Show*, magazines such as *Maxim*, and Coors Light beer commercials embody and valorize this popular form of manhood.

Sports talk radio, linked with the other masculinist genres previously noted above may represent an attempt to symbolically reassert straight men's superiority over women and gay men (Smith, 2002). In this vein, David Theo Goldberg (1998) suggests that sports talk radio, far from being a democratizing force (here disagreeing with Haag), reinscribes dominant discourses and is a leading forum for reproducing male domination. He contends that "sports talk radio facilitates this masculine self-elevation, the ideological reproduction of hegemony—risk and cost free but for the price of the toll call" (p. 218). Smith (2002) suggests that sports talk radio is an audio locker room that reinforces hegemonic masculinity and suggests that the locker room is a key site of male privilege and a center of fraternal bonding (p. 1). She writes:

> Men attach deeply personal meanings to "being a sports fan." Sport talk radio shows have been able to capitalize on utilizing the airwaves to create "communities" despite the physical distance between listeners and from the host and still provide that emotional attachment that fans seem to search for. Unlike a television, which would be difficult to transport around and requires consistent visual attention, the radio requires the listener to hear, making multitasking possible. Listeners can participate as they drive to work, sit in their cubicle, deliver packages, exercise, or sit at home. (p. 8)

Self-confessed addict of sports talk Alan Eisenstock (2001) wrote a book titled *Sports Talk*, a masculinist celebration of the significance of sports radio and the sports talk radio junkie. He refers to sports talk shows as a "non-stop fraternity party, a sport bar on the radio" (p. 3), in which men, through the medium of a call-in program, can interact with other men free from the censure of feminism and political correctness. Sports talk radio, from this perspective, is a mass-mediated attempt at preserving male-only spaces reminiscent of the rise of fraternities and the Boy Scouts around the turn of the twentieth century (Kimmel, 1996).

As noted earlier, homosocial spaces became popular once again beginning in the 1960s with men who were interested in addressing and changing masculinity. Judith Newton (2005) argues that men's movements, albeit diverse and contradictory with differing agendas (profeminist, men's rights, Promise Keepers, and the mythopoetic movement), share one element in common: "male romance." Newton refers to "male romance" as an effort to transform masculine ideals by going off with other men in a homosocial space to enact particular rituals that provide a sense of being "born again." Feminists have long criticized male romance by suggesting that it almost always works to reinforce white, middle-class, and heterosexual male power. Daniel Lefkowitz (1996) suggests that the popularity of sports talk shows depends on "the same cultural dynamic that lends dynamism to the Men's Movement" (p. 210). The appeal of sports radio, according to Lefkowitz, depends in part on on this notion of male romance—namely, a desire to engage in homosocial bonding via the ritual of sports fandom. Hence, sports radio could be viewed as a commodified, mass-mediated version of male romance.

This book will critically examine the link between the cultural phenomenon of sports talk radio and organized efforts by men to reinvent masculinity (sometimes referred to as men's movements). I will inquire if sports talk radio, as part of the "frat boy nation," shares the goals and values of some men's movements. Does the discourse of sport talk radio consolidate male dominance and reestablish traditional gender relations and roles? Is sports radio just a crude marketing device that affirms hegemonic masculinity without confronting or questioning it? Can sports radio, while highly market-driven, open up some space to transform masculinity? I will attempt to complicate Goldberg's (1998) assertion that sports radio uniformly reinscribes dominant positions of power by exploring the ways the genre may allow some promise to reinvent masculinity as it simultaneously reinforces traditional gender relations. This book will also extend beyond the few academic articles that have been written on sports radio that have focused solely on textual analysis (Goldberg, 1998; Haag, 1996; Mariscal, 1999; Smith, 2002; Tremblay and Tremblay, 2001). *Beer, Babes, and Balls* includes interviews and fieldwork with actual fans of sports radio. This opens up space for more complex analysis in terms of how the text interacts with consumption.

MEDIA AND CULTURAL STUDIES

This book's inquiry into sports talk radio is informed by cultural studies theories and methodologies. I am aligned with cultural studies work that has called into question assumed hierarchies of "high" and "low" culture by turning critical attention to formerly disparaged media forms such as women's magazines, working-class style, popular music, romance novels, and television (Ang, 1996; Grindstaff, 2002; Hall and Jefferson, 1983; McRobbie, 1991; Morley, 1992; Radway, 1984). This strand within cultural studies is in contrast to certain traditions within critical theory—namely, the Frankfurt School. Frankfurt School theorists Theodor Adorno and Max Horkheimer (1997) argue that cultural products/texts are commodities produced by the culture industries that, while purporting to be democratic, are in actuality conformist and authoritarian. I believe that the Frankfurt School's analysis, while very useful and compelling, holds to an overly monolithic view of the culture industries (media organizations that produce and distribute art, entertainment, and information), and denies the capability of consumers/audiences to be active producers of meaning rather than passive victims.

Cultural studies draws from the fields of anthropology, sociology, gender studies, feminism, literary criticism, history, and psychoanalysis in order to examine contemporary media texts and cultural practices. It has broadened beyond the sphere of a sole focus on political economy (studying the production end of the cultural industries) and texts (anything that produces meaning) to encompass a focus on audience reception and meaning-making. Research within this cultural studies tradition takes as its starting point a belief that media texts cannot be examined in the abstract; instead, what is crucially important is how audiences respond to texts.

Beers, Babes, and Balls draws upon the cultural studies tradition of simultaneously studying the production, textual content, and reception of the mass media. Media scholars argue that in order to understand a cultural phenomenon, one must understand the interrelationship between the activities through which the text is produced, the messages in the text, and how those who consume the text interpret it (Davis, 1997; Gamson, 1998; Jackson, Stevenson, and Brooks, 2001). Studying the production, text, and consumption aspects of sports talk radio requires me to engage in a variety of research methods in different contexts. In addition, these diverse methods and locations produce potentially different ethical and methodological problems. This book discusses some of the ethical, epistemological, and methodological dilemmas and questions that arise in conducting a triangulated analysis. For instance, one of the most challenging parts of the book was the audience response component. Accessing the audience was more difficult than my reading of studies of the television audience had led me to expect. A reflexive account of the research process will allow me to illustrate some of the difficulties of doing audience response.

There is a growing body of sporting analysis informed by cultural studies scholars (Birrell, 1988; Birrell and McDonald, 2000; Butler, 1990; Cole, 1993; Dunning, 1986; Messner 1992). These scholars are engaged with the intersection between feminism and cultural studies. Feminist cultural studies is based on the assumption that power is distributed inequitably throughout society, often along lines of gender, class, race, and sexuality. These relations of power are not fixed, but contested. Moreover, power usually is not maintained by force but through more subtle forms of ideological dominance (Gramscian hegemony theory). Ideology is the set of ideas that serve the interests of dominant groups, but are taken up as the societal commonsensical even by those who are disempowered by them. Sport is a particularly public site for such ideological struggle: "What is being contested is the construction and meaning of gender relations" (Birrell and Theberge, 1994). The utility of the theoretical vocabulary of cultural studies to explore the intersections of gender, race, and class in sport has been clearly recognized. It is this struggle that interests me and other critical sport scholars.

CRITICAL RADIO STUDIES

This book also draws upon critical radio studies. In its heyday—the 1920s, 1930s, and 1940s—radio occupied an exclusive position as the only home-based electronic mass medium (Cook, 2000; Hutchby, 1996). Radio scholar Michele Hilmes (2002) states that "radio provided one of our primary means of negotiating the boundaries between public life and the private home, becoming the American family's 'electronic hearth' and our central acculturating and nationalizing influence" (p. 1). Abandoned by the media networks for television in the late 1940s, radio plays a diminished role in the United States. With the ensuing dominance of television, radio became the poor relation in media and cultural studies. The recent radio scholarship, informed by cultural studies, argues that radio continues to be an important cultural form, problematizing the distinctions between public and private and raising questions about the primacy of voice and sound as a central and potentially subversive feature of subjectivity.

This book thus seeks to study sports talk radio by situating it within its historical and institutional context. To achieve this goal, this book explores some of the previous studies on talk radio. Much of the research on talk radio emerged in the 1990s, exploring its potential democratic functions. Writers placed great emphasis on the significance of the opportunity provided for audiences to participate in mass-mediated debate and discussion. Researchers have focused on the role talk radio plays in keeping listeners up-to-date with political issues, and how talk radio shows provide a forum where these issues can be discussed by ordinary citizens (Page and Tannenbaum, 1996). The consensus here is that this participation has positive functions, both for individual callers and for the democratic system.

A serious shortcoming of these studies of talk radio is that they tend to overlook the fact that the majority of radio stations are commercial broadcasters competing for advertising revenue that is attracted according to niche demographics and audience size (Hilmes, 2002). The most important of these is profit-making via advertising. The commercialization of radio casts serious doubt on the potential for talk radio to be a democratic forum. Talk radio is not unique in this respect as Hilmes (2002) identifies a general trend whereby the commercial functions of the U.S. media are becoming increasingly important. This leads to consumption taking over from participation, and the audience increasingly being positioned as consumers rather than as citizens. Hilmes also notes that the Telecommunications Act of 1996 removed the barriers to ownership of multiple stations in the same market, provoking a wave of station purchasing and media consolidation of territory.

Only two studies have examined talk radio as it is understood by its audiences (Herbst, 1995; O'Sullivan, 1997). This is not surprising. As stated earlier, the radio medium has been neglected by media studies scholars in recent decades and has taken second place to television. There have been a just a few empirical studies of the medium (e.g., see Moores, 1993). However, almost no attempt has been made to theorize the genre, with the exception of Erving Goffman (1981). The result of this neglect is that turning on the radio is seen as something "natural," something that is done by most people every day:

Radio, in this age of television predominance, has taken on the role of a familiar family member—accepted, unquestioned and treated as part of the scene. Popular commentators and researchers alike have focused our attention on the electronic tube, to the neglect of radio. Radio, however, continues to outdraw audiences in both time and number. It is . . . an important part of the cultural day. (Moss and Higgins, 1982, p. 282)

In contrast, work focusing on television, and in particular on the television audience, has been both plentiful and theoretically rich. Hilmes (2002), in addressing the future direction of radio studies, suggests that a "greater attention to audience and meaning making from a cultural studies perspective could help to bring radio into the mainstream of academic study and provide a necessary and provocative corollary to the many important findings in the area of television studies" (pp. 13–14). This book is a response to Hilmes's call as it fills a much-needed gap in the area of radio studies. It will also move beyond the radio studies' focus on talk radio's relationship to democracy and politics. As Douglas (2002) argues, talk radio plays a central role in reestablishing male privilege: "talk radio is as much—maybe even more—about gender politics at the end of the century [twentieth] than it is about party politics" (p. 485). This book examines the relationship between sports talk radio and male privilege.

OUTLINE OF THE BOOK

Beers, Babes, and Balls is divided into three parts. Part I—The Climate for Sports Talk Radio—will examine the production aspect of sports radio. It will first sketch the history and impact of talk radio from a political economy perspective, which means studying the interconnections between corporate ownership, the radio industry, and sports radio genres. Rather than a strictly linear approach to production–content–audience, this book will explore the connections between post-Fordist economics (the emergence of flexible specialization), culture industries (niche marketing), and society (shifting gender and sexual relations). This section will also include interviews I conducted with various producers and hosts of both local and national sports radio programs along with my brief participant observation of the production of a local sports radio show. I will highlight some of the tensions and contradictions that are part of the sports radio production process.

Part II—Reading Sports Talk Radio—provides a textual analysis of the nationally syndicated sports radio program *The Jim Rome Show.*

I have selected this program because it is the most popular national show and is fairly representative of the genre of sports talk radio in general. I taped the show on an ongoing basis and downloaded transcripts of programs from Jim Rome's Web site (www.jimrome.com). I am particularly interested in analyzing instances on the program that stand out as important moments, what journalists often call "pegs"—critical events that generate a flurry of coverage (Grindstaff, 1994). I am chiefly concerned with pegs that focus on issues of sex, class, gender, and race. Using these themes, this section analyzes the ways the show serves to both reinforce and challenge hegemonic masculinity.

To help work against the limitations of critiquing texts in isolation from context, Part III—The Audience of Sports Talk Radio—is an ethnographic account of the sports radio audience to better understand the meanings and uses of sports talk radio in the everyday practices of living by some of its fans. I have chosen to conduct this fieldwork in sports bars because many of the patrons who frequent these spaces are avid listeners of *The Jim Rome Show* and other sports radio programs. In addition, since it is a primary site for male bonding, the sports bar is an extension of the social practices and discourses evident in sports talk radio (Wenner, 1998b). Given that my research will be limited to a small number of participants and because the audience members I interview may not be representative of the North American sports radio audience, the results are not necessarily generalizable. Yet my hope is that my findings will promote insight into future research on the ways that listeners decode sports talk radio texts. I am particularly interested in exploring the pleasures associated with listening to sports radio, the imagined community that is created through sports radio, and the meanings that listeners make of some of the more progressive moments of *The Jim Rome Show*.

The concluding chapter summarizes the connections between the production, text, and consumption of sports talk radio. It examines the messages that circulate on sports radio, its listeners, and the larger media and societal dialogue on masculinity and gender relationships. This chapter will also summarize the ways that sports talk radio serves as an important mediated site for male bonding, helping men feel empowered in a society in which the gender order is changing. It will make some conclusions about how male bonding in sport talk radio not only reinforces hegemonic masculinity but may offer some potential for men to alter traditional manhood.

PART I

THE CLIMATE FOR SPORTS TALK RADIO

2   2nd Inning

THE SPORTS TALK RADIO INDUSTRY
From Rush to Rome

I wake up to sports talk radio. When I am in my car, I listen to sports talk radio. I often go to sleep listening to sports talk. It's a comfort and pleasure for me to hear familiar voices discussing the day in sports and it helps reduce the stress of the day. Rarely do I think of the production or corporate end of sports talk radio. When I am listening in my car to sports talk, the process of production—the screening of calls, the control of advertisers over the content of the discussion, the ratings pressures, and the corporate influences—only occasionally reveal themselves; I mostly become absorbed in the so-called literal discussion of sports and which team won last night. Sports talk radio presents its format to me (and to other listeners) as a "democratic" and highly participatory format that reflects a free marketplace of ideas—in this case, sports discussions.

But how does sports radio work behind the scenes? In sports talk radio, who produces the speech? Is it journalism and news? Entertainment? Natural? Constricted? What is the history of sports talk radio? How did it become so popular? What were the economic and political forces behind its growth? Is it a democratic forum or a friend of excessive corporate power? To answer these questions, I first examined the history of talk radio and deregulation, and how it grew in the 1980s due to several economic and social factors. Historical analysis helps provide a background into sports talk radio's growth, its link to the larger talk radio industry, the current global economy, and its relationship to hegemonic masculinity.

RADIO DEREGULATION AND TALK RADIO

The rise of sports talk radio was a cultural phenomenon of the late 1980s and 1990s that depended a great deal on the restructuring of the AM band and the eruption of talk radio as a low-cost programming alternative (Mariscal, 1999). Talk radio began to make national headlines in the mid-1980s, when Howard Stern gained increasing notoriety, earning the title of "shock jock," and Alan Berg, an especially combative talk show host in Denver, was murdered, apparently by an irate caller.[1] More headlines came in 1989, when a coalition of some thirty talk show hosts organized a major attack on a proposed 51 percent congressional pay increase that then Speaker of the House Jim Wright planned to push through without a floor vote.

The number of radio stations with a talk format mushroomed from 400 in 1987 to 900 in 1994 to 1,200 in 2001 (Douglas, 2002). Music formats had abandoned AM for the superior quality of FM. Except for a handful of AM "full-service" stations, the perceived market value of AM stations dropped quickly. Since talk radio did not require stereo of FM fidelity, and it was unpredictable and participatory, the format proved to be a solution to AM's desertion.

Another key ingredient to fueling the popularity of talk radio was the cell phone. Virtually unheard of as an automobile accessory in the early 1980s, cell phone sales exploded between 1989 and 1992. During that period, the number of subscribers to cell phone services increased by 215 percent; by 1995, there were 33 million subscribers (Douglas, 2002). And one of the things that these cell phone subscribers did as they drove to and from work was listen to and call talk radio shows. "Those trapped in traffic jams are captive audiences for radio," observes Haag (1996). "All you have to do is look at the freeways at quarter to five. Every traffic jam is an opportunity to a broadcaster. You're dealing with a medium that has very much a captive audience in the automobile. It's bordering on the oppressive" (p. 456).

Talk radio's appeal appeared to tap into the sense of public life, the isolation and exhaustion that come from overworking, and the increasing gap people felt between themselves and politicians. The genre represented a novel and often brash and aggressive way of creating a group identity within the homogenizing blitz of conventional media fare. Most of the critical commentary about talk radio focused on two aspects: the threat it posed to civility due to its offensiveness, and the threat it posed to democracy due to the rise of right-wing talk show hosts. What has been less discussed is talk radio's pivotal role in efforts to restore male privilege to where it was before the second-wave feminist movement. After all, over 80 percent of the hosts and a majority of listeners are male. Susan Douglas (1999) writes:

> There were different masculinities enacted on radio, from Howard Stern to Rush Limbaugh, but they were all about challenging and overthrowing, if possible, that most revolutionary of social movements, feminism. They were also about challenging buttoned-down, upper-class, corporate versions of masculinity that excluded many men from access to power. (pp. 289–290)

From the start, the talk on talk radio was "decidedly macho and loud" (Douglas, 2002, p. 488). Talk radio hosts' aggressive speech helped build imagined communities that made clear who was included and who was excluded; there was no space for men who were too sympathetic to the civil rights or women's movement—these men were mama's boys and wimps. Hosts insulted listeners like abusive fathers but tough callers knew how to take it. In fact, talk radio proved to be a decidedly white, male preserve in a decade when it was much less permissible to offend women, gays, lesbians, and the poor—the very people who challenged heterosexual, white male privilege. Now, with talk radio, it's payback.[2]

As scholars Susan Jeffords (1993) and Michael Kimmel (1996) have noted, the 1970s was a period of anxiety about manhood in America. In fact, some have argued that this was a true moment of crisis for white masculinity. Feminists had made gender politics headline news, and demonstrated how patriarchy undermined and threatened democracy and equal opportunity for all. America's patriarchal power structure was affronted as Americans were held hostage by Iran. American presidents, particularly Jimmy Carter, had lost control, and control and mastery are central to most conceptions of traditional manhood and American national identity.

Ronald Reagan, through his rhetoric, politics, and appearance, sought to change all that. "Screw feminist politics and getting in touch with your feminine side, said the Reagan presidency. All that had done was make the country vulnerable, flaccid, and weak. It was time to reassert male supremacy" (Douglas, 2002, p. 489). Hollywood responded to Reaganism with high-action movies in which Sylvester Stallone, Arnold Schwarzenegger, Tom Cruise, Bruce Willis, and others used their tough, chiseled, masculine bodies to remasculinize the American image, which fit very well with Reagan's efforts to pump a great deal of testosterone into American foreign policy, the fight against crime, and the "war on drugs."

On talk radio, the trend was similar—to take over public discourse, purge it of feminine tendencies, and reclaim it for men. In *Talk Radio and the American Dream*, Murray Levin taped seven hundred hours of talk radio and found among callers a discourse worried about emasculation. The natural order of things now seemed reversed, so that crime, blacks, rich corporations, and women all had the upper hand. Talk radio became the discursive battleground on which to reclaim hegemonic masculinity and rid the United States of soft-spoken, New Age guys. Even though the callers lacked the power to ward off the verbal put-downs of the host, they kept coming back for more.

The participatory format of talk radio, along with its suggestion that it would reverse years of the ongoing consolidation of power by corporations and Washington, held its central appeal. The great irony is that this very kind of talk radio, with its macho version of populism, was the product of deregulation, merger mania, and corporate consolidation during the 1980s and beyond. Populism and civic involvement were the public faces of radio; they disguised increased economic concentration and heightened barriers to entry for all but the very rich and powerful into the industry itself. As Douglas poignantly suggests, "that was the Reagan administration's great genius—selling the increased concentration of wealth as a move back toward democracy" (p. 491).

While gender politics in the current global economy may, on the surface, appear gender-neutral, it does have implicit gender politics. For instance, the attack on the welfare state weakens the status of women, while the unregulated power of global corporations (including radio conglomerates) places power in the hands of specific groups of men (Connell, 2000). Sports radio is owned by corporations and operated by conservative men who subscribe to hegemonic forms of masculinity. Likewise, the hosts and production staff, with a few exceptions, are men who engage in masculinist sports talk that may be part of a feminist backlash.

The director of the Federal Communication Commission (FCC) under Reagan, Mark Fowler, championed the deregulation of radio in the 1980s, allowing companies to own greater numbers of stations and eliminating restrictions on how long a company had to own a station before reselling it for a higher price. The other significant deregulatory move in the 1980s was the abandonment of the Fairness Doctrine, which the FCC announced it would no longer enforce. This doctrine required stations to offer access to air alternative opinions when controversial issues were discussed. The goal of the doctrine was to promote a balance of views. Opponents of the doctrine, including Fowler and Reagan, felt it inhibited freedom of speech. Stations, they argued, avoided giving air time to opinionated individuals because of the requirement to broadcast competing points of view. Unrestricted by the Fairness Doctrine's mandate for balance, Limbaugh and a legion of ultraconservative imitators took off the gloves and revived the financial state of AM radio. WFAN, the first all-sports station in the United States, premiered in 1987, the same year as the revocation of the Fairness Doctrine.

The Telecommunications Act of 1996 furthered merger mania by significantly loosening ownership restriction on broadcast stations, removing all limits on the number of stations a given company can own nationally, and raising from four to eight the maximum number of stations that a company can own in a given market. The 1996 Act thus touched off an onslaught of massive consolidation within the radio industry. Within a year of the act's passage, huge radio oligopolies were created such as when Clear Channel Communications expanded its ownership to 80 stations and Viacom bought Infinity Communications to form a network of 77 radio stations (Douglas, 1999). Currently Viacom/Infinity owns 180 stations, including many sports talk radio stations. Premiere Radio Networks, a subsidiary of Clear Channel, currently syndicates 70 radio programs and services to more than 7,800 radio affiliations and reaches over 180 million listeners weekly. Premiere Radio is the number one radio network in the country and features Rush Limbaugh, Dr. Laura Schlessinger, and Jim Rome, among others.[3]

Such overwhelming consolidation has resulted in the centralized operation and management of stations, such that major programming decisions are made by national rather than local offices. The end result for listeners and communities is that radio stations fail to reflect their local communities very well, giving short shrift to local news and sports programming. Various political economists, including Robert McChesney (1999), are increasingly concerned with the threat to democracy and civic discourse posed by such rapid consolidation of corporate concentrated ownership of U.S. radio stations.

SPORTS TALK RADIO:
AN EXTENSION OF POLITICAL TALK RADIO?

Corporate consolidation in the media industry developed alongside significant economic changes: a move from a manufacturing economy to a service economy accompanied by increased privatization, a weakened state, and more multiple, specialized markets. Economic success is now determined by product innovation, flexible specialization, and niche marketing. Sports radio, according to Haag (1996) is "niche marketing at its most powerful . . . very few women, very few kids" (p. 459). Advertisers can count on a well-defined niche to promote men's clothing, cell phones, automobile products, and alcohol. In an interview in a recent issue of *Forbes*, Clear Channel's CEO, Lowry Mays, was quoted as saying, "If anyone said we were in the radio business, it wouldn't be someone from our company. We're not in the business of providing news and information. We're not in the business of providing well-researched music. We're not in the business of talking sports. We're simply in the business of selling our customers products" (cited in Solomon, 2003, p. 1).

Sports talk radio, like talk radio in general, continues to grow with the current deregulatory trends in the industry. In 1999, there were 150 sports radio stations (Mariscal, 1999); now there are 348. WFAN in New York is currently the most profitable radio station in the world, bringing in over $50 million in advertisement billing from 2000 to 2003. According to Arbitron, the radio industry ratings and marketing service, the niche for sports radio is white men, ages twenty-five to fifty-five, who listen on average to sports radio six hours per week (75 percent of listeners are Caucasian, 11 percent are black, 6 percent are Asian, and 7 percent are Hispanic). Arbitron's research also states that half of the sports audience is in the highest income bracket ($75,000 annual income). Hence, Arbitron, in a Power Point presentation to sports radio station managers, encourages stations to think about out-of-home marketing and products since upper-income men spent most of their time on the road commuting (Arbitron also states that 60 percent of all sports listening occurs in the car).[4] Arbitron's research confirms that sports radio's audience is the same as political talk radio's niche.

Ceding the radio airwaves to niche marketing and global capitalism has predictably generated programming that mirrors the politics of those who champion right-wing policies. Academic Maureen Smith

(2002) writes, "Where Rush Limbaugh gave politics another bully pulpit and reached an audience that was not subscribing to *The National Review*, sports talk radio shows have mimicked the style and successes of talk radio." According to scholars Goldberg (1998), Haag (1996), and Mariscal (1999), sports talk radio is just as hostile to feminists and gays as political talk radio. In fact, according to Kevin Cook (1993), sports talk radio, even more than political talk radio, is the only arena left for white men who have been "wounded by the indignities of feminism, affirmative action, and other groups' quest for social equality" (p. 20). In line with Cook's argument, Haag states:

> Sports talk show is a venue for the embattled White male seeking recreational repose; that it caters to this audience as surely as Rush Limbaugh articulates its discontents. Some sports talk stations define their listening audience explicitly as the Atlanta sports station [The FAN] manager states, "we make no pretensions about what we're doing here. The FAN is a guy's radio station. We're aiming at the men's bracket which is the hardest to reach. (p. 459)

So, is sports talk radio just an antidemocratic forum that undermines social justice? It is merely a sexist medium? If that is the case, why do I, a feminist scholar who is concerned with corporate greed, listen on a daily basis? In my experience, sports talk radio, unlike right-wing political talk radio, is more tolerant of diverse perspectives and there are times when issues such as gender and race are substantively discussed—it is not simply a feminist backlash medium (see chapters 5 and 6). To further examine these questions, I interviewed several staff inside sports talk radio to understand how they experience the industrial, commercial, and gendered imperatives under which they operate. How do people in the sports talk industry relate to the norms and values of corporate power and male hegemony? These issues are addressed in the next chapter.

3   3rd Inning

INSIDE THE SPORTS RADIO INDUSTRY
Ads and Lads

This chapter is based on a series of interviews with producers and hosts of both national and local sports talk programs along with participant observation research at a local sports talk radio station. Interviewees included a national sports talk radio host, two local sports talk hosts, a sports talk radio station manager, and a producer of a local sports talk show (the only woman I interviewed). I interviewed these participants via email or a phone call to the radio station. All of the interviewees preferred to remain anonymous, with the exception of Kevin Wheeler, the host of the Sporting News Radio's nationally syndicated *Kevin Wheeler Show,* which airs weekdays from 2 A.M. to 6 A.M. ET.[1] The interviews were semistructured, guided by a set of predetermined questions with a number of branching questions that were used to elicit more detail and more focused attention to the study's domain of interest. Probing questions were also used spontaneously to prompt elaboration and specificity.

I asked four major questions during the interviews: (1) Describe how you got involved in sports talk radio; (2) Why do you think sports talk radio is so popular with men?; (3) Describe your show's target audience; (4) What are some of the pressures in the sports talk radio industry? I asked these specific questions because I wanted to find out from sports radio staff how they experience the industry from the inside, particularly the commercial pressures, and sports talk radio's relationship to traditional masculinity.

I realize that although my interviews may contribute some valuable information about their roles in the sports talk radio, this research method is not without limitations. It has been my experience that people in the media industry are wary of academics; they often believe that scholars read too much into the messages in the media. People in the media industries think their production practices are normal and ordinary—they take for granted what they do and say at work. I did not expect anyone to necessarily "spill the beans," since most industry staff will not likely critique the negative sides of the sports business to an outsider, particularly an assistant professor. Hence, I attempted to anticipate what their possible responses were going to be and whether this would be helpful in answering my research questions. My experience as a clinical social worker/therapist was helpful in this process. My skills at narrative therapy interviewing—building trust and developing a context of safety that fosters an alliance—helped me attend to those "unique outcomes" in the interviews—moments in the conversation that were unexpected or surprising.

The participants determined the settings for the interviews. The location for the interviews with one of the local hosts was the radio station. The interviews with the producer and Kevin Wheeler were conducted via email. My discussion with the station manager and the other local host occurred on the phone (via speaker phone so I could tape the interview). The data analysis began with a verbatim transcription of the audio-recorded and email interviews. Once this was completed, a qualitative content analysis was conducted to examine salient topics covered, patterns, regularities, and differences within and across the subjects. I was able to identify three organized themes that emerged from the transcribed texts: (1) ambivalent attitudes toward the corporatization of radio; (2) sports talk radio providing a space for public dialogue; and (3) a nostalgic and romantic belief in sports. These three themes will be analyzed in detail in the following sections.

INFLUENCE OF ADVERTISING, RATINGS, AND CORPORATE RADIO

While driving on the freeway recently, I noticed a billboard that said, "Armstrong and Getty—Listen to them before we fire them." Armstrong and Getty are local Sacramento talk show hosts who have quite a popular following. However, on their show they frequently mention the fear of being fired for saying something offensive or defying their station manager. The billboard is reflective of the volatility of the radio industry. There is a long history of hosts, disc jockeys, and program directors getting fired, moving around to different cities and different stations. Howard Stern's autobiographical movie, *Private*

Parts, chronicles the famous "shock jock's" firings and rehirings due to offending station managers and breaking FCC rules.

My analysis of the production staff interviews reflects this high level of job turnover and career insecurity. Both hosts and producers talked about sports radio industry reality: never knowing where you might end up and never knowing when your contract might not be renewed because of poor ratings. The station manager said, "We are all just renting time in radio; our jobs are never safe." In fact, while I was interviewing some local hosts and producers at one particular station, one of their colleagues (the host of a morning show) was fired by the station and the corporation (Infinity) due to insufficient sports knowledge and lackluster ratings. One host said that ratings are a "constant source of tension—you never stop thinking about it. And it's so fleeting; one month you're up and the next you're down." Several said they take the ratings personally. The Arbitron ratings, according to the station manager I interviewed, "determine the advertising rates; good monthly rating means a higher rate. A bad month of ratings is torturous and a good month means I can sleep better." A program director interviewed by Alan Eisenstock (2001) described the pressure in this way:

> As a program director, you constantly worry. You get ratings every month. I have friends who are program directors who literally throw up every month. I don't do that. I don't take it that much to heart. The way I look at it, when the ratings are good, they're never good enough. And when they're bad, I take it personally. My job . . . I have to be the number one in my target demographic. (p. 165)

Many spoke of the multiple balancing acts that sports radio work involves, including the pressure to attract advertising revenue while staying loyal to callers. All my interviewees referred to advertising as "a necessary evil." Attracting advertising revenue is a constant source of tension for many, particularly for the station manager I interviewed. After making sure that I was going to honor his anonymity, he was openly critical of the advertising industry:

> I do recognize their power . . . They [the advertisers] are our number one priority . . . let's face it . . . without them we can't give our listeners the sports stuff they want. But it is hard to always push new products like the latest gadget or male enhancement pills! I got into this business because I love sports and call-in programming, not to push products.

Sports talk radio stations have also been criticized by media watchdog groups for their lack of journalistic credibility and impartiality. Many local sports hosts throughout the United States are also employed by major sports team franchises to broadcast games. For instance, KHTK, the sports talk radio station in Sacramento, California, has a contract with the NBA Sacramento Kings to air their games. KHTK's sports talk hosts are also employed by the Kings (the Kings are owned by Maloof Sports & Entertainment) to do play-by-play and pre-game and post-game commentary. Jeff Kearns (2003), in his article "Embedded with the Kings," suggests that the KHTK's hosts are "cogs in an expansive promotional and media machine that seemingly mixes Kings announcers, players, media outlets, and advertisers—all of whom capitalize on and profit from the success of the only big-name sports team in town" (p. 1).

Kearns's critical exposé alerts us to the increasing synergy between major league sports franchises, corporate sponsors, and media outlets. These alliances call into question the notion of journalistic independence; as Kearns writes, if "news organizations aren't independent, can they fairly and aggressively cover non-sports stories about sports franchises to which they're tied" (p. 2).[2] By creating partnerships with sports radio stations, sports franchises are able to use the station to advertise their products and construct a positive identity for the public. At one Kings game I attended, it was apparent that the promotional campaign is successful; many fans strolled through the Kings gift shop, browsing through $50 jerseys and $350 jackets emblazoned with the team logo. Inside the sold-out arena, the King's corporate alliances were posted everywhere—Cingular, Folsom Auto Mall, KHTK, Channel 10 News, and the *Sacramento Bee*.

When Kearns interviewed Doug Harvill, market general manager of Infinity, which owns KHTK, he denied that the Kings get any special treatment with editorial coverage. Similarly, the hosts I interviewed, while acknowledging the increasing trend of partnerships with teams and sports radio stations, feel no ethical dilemma, vehemently denying any pressure from the teams they cover. However, the Sacramento female producer (whom I call Beth), contradicting her peers, feels pressure from the Kings and other corporate sponsors:

> You just know that you aren't supposed to badmouth the people (the sports franchise owners, advertisers) who pay you . . . We do have to kiss ass to the advertisers, the Kings, and the corporate sponsors all the time . . . politics. It's bullshit. If we are supposed to have journalistic freedom, we should be able to rip a coach/player/organization without them or their sponsors being upset about what you said.

The tension between journalistic independence and the necessity to maximize advertising revenues may be a sign of the paradoxes and ambivalences of current masculinities as well as a reflection of the dynamics of commercial culture. Many of the products advertised on sports radio—automobiles, beer, gadgets, and male enhancement pills—are reflective of the laddish masculinity mentioned earlier (Benyon, 2002). Similarly, station ads (such as "sports talk radio—it's just beer, babes, and brats!") give the impression that the staff are not really working at all; it's just one big fraternity party. Yet, all the hosts and staff I interviewed talk about long hours, fatigue, and work stress. How might we understand the discourses of hedonism in face of increased corporate pressures and work strain? One argument might be that the emphasis on pleasure-seeking is assembled to mask the increasingly bureaucratic and rational features of the modern workplace. Stories of sports radio as one big laddish celebration obscure the fact that sports radio staff are all involved in rational bureaucratic work organizations—a feature of many men's work experience in today's hypercapitalist culture (Faludi, 1999).

When asked about the direction and trends of the sports radio industry, the two local hosts were quite critical of the corporatization and media conglomeration of radio. Both resented the fact that they felt remote from important decisions, which were increasingly made by program directors, station managers (referred to as "monkeys"), and corporate headquarters. "They just want numbers; they don't care," said one host. The other host expressed concern about the syndication trend, stating that sports radio was losing the localism and regionalism that have historically been such a part of sports:

> I don't like small, local stations being "gobbled" up by big media conglomerates. Baseball is a local sport; you lose the nuances and subtleties on a nationally syndicated show . . . Plus, when small stations get gobbled up, it is replaced by some guy at corporate who doesn't know the community . . . It then is just about pleasing the sponsors and niche marketing . . . it's about saving money. I don't want a national show. I like staying connected to a community.

Kevin Wheeler, the national host I interviewed, had quite a different perspective. When I asked him about his thoughts on the radio industry, he replied, "No comments, really. Sports radio is just like any other industry in that it is constantly evolving." Wheeler's response is not surprising; he works as a nationally syndicated host for a big media company. Criticizing the trends of media conglomeration would potentially put his career at risk. Furthermore, Wheeler's career has benefited from corporate trends and syndication. Wheeler's comments are also representative of many people who work in the corporate sector and have accepted the trends of a global economy. Neoliberalism has triumphed—democracy is equated with capitalism—better to join the economic trend than fight is the common mantra.

Lastly, the station manager had some interesting remarks about radio deregulation and recent events with the FCC and Congress. In June 2002, the FCC voted to ease media ownership limits even further. This FCC vote produced a strong backlash response by various media activists groups (such as the Internet-based group, Media Reform Network) and by Congress, who expressed concerns about corporate interests' taking precedence over diversity, localism, and democracy. On September 16, 2003, the Senate, wielding a rarely used legislative tool, repealed the new FCC rules relaxing media ownership regulations in a bipartisan vote. The convincing 55–40 vote was a setback for the Bush administration and Senate Republican leaders, who backed the new regulations. The station manager felt positive about the Senate vote, saying,

> However, citizens and the Congress have every right to look at media conglomeration. Since deregulation, the radio industry has changed, not all for the good. Many radio stations have lost their local, "community" feel to it. With the congressional hearings, however, it has influenced the industry to make changes and get back to the local. This backlash is good, I think. In fact, lately, my boss at corporate [Infinity] has been encouraging me to do more local programming—high school games, the like. We were not doing local stuff a year ago. It's a good thing and wouldn't have happened without the Congress and citizens protesting the FCC rules.

His comments suggest that various citizen-based groups are effectively organizing and changing the direction of media programming to become more democratic, more diverse, local, and participatory. Only time will tell if this current countermovement takes hold and reverses the direction of media consolidation.

SPORTS RADIO AND PUBLIC DISCOURSE

Susan Douglas (1999) posits that people turned to talk radio in response to a breakdown of public life (that has occurred since the 1980s). Talk radio, according to Douglas, fulfills a timeless need people have to be involved in civic affairs and express their opinions with fellow citizens:

> Talk radio gave listeners a way to tap into the nation, into public opinion, into a community that they did not have before, where they could hear viewpoints that had not been filtered and homogenized by the TV networks and their news anchors . . . Listeners find themselves politically isolated at work or at home, deprived of any forum for discussion or debate. Co-workers and family members were either politically apathetic and ignorant or of a different political persuasion, which meant that going back and forth with them about current affairs would be frustrating, even infuriating. But tuning into talk radio, people could hear other points of view, even outrageous points of view, and they could take them in quietly, or scream back at the radio without fear of an altercation. (pp. 311–312)

Clearly, talk radio speaks to many people's need to have a voice and hear others as it has tapped into the alienation experienced by the increasing privatization of American life.

The hosts I interviewed said something similar when I asked why sports radio is popular:

> *Kevin Wheeler:* Sports radio is popular for the same reason regular talk radio is popular—people feel like that they have a voice. Even if an individual doesn't get to call, they know that there are others out there like them who will . . . Sports radio is about interaction, in my opinion. Callers take the time to call (and hold for thirty–sixty minutes sometimes) because they want to be heard, even if the expression of their opinions doesn't effect a change.

> *Local host:* Why is it popular? Because guys like to talk sports . . . it's in our genes! We like to mix it up with other men; feel heard and express our opinions. We used to do that at bars, in our neighborhoods . . . But now, we are working all the time, so we do it in the car . . . It's how we connect.

Both Wheeler and the local host's comments resonate with the idea that in late capitalist and privatizing culture, sports radio attempts to satisfy a need for humans (in this case, men) to participate in the public realm. Susan Herbst (1995) refers to this civic engagement as an "imagined community" created in electronic public space. Since many men in a neoliberal economy are working and living increasingly isolated lives, sports talk radio gives the listeners and callers a discursive space to create community and enjoy social interaction.

Likewise, Pamela Haag (1996) believes that sports radio fulfills people's desires to be "thrown together in unexpected, impassioned, even random social relations and communities" (p. 467). Haag concludes that it fashions civic discourse in public space and acts as an alternative to "hate" political radio. In addition, she suggests that the dialogue in sports radio, mostly civil and respectful in tone, reinstates an older form of regionalism that "concretizes social rather than economic communities" (p. 460). In this sense, Haag contends that sports radio promotes local discourse in the face of commercialized blandness, creating a fissure in the hegemonic "corporate voice" of the mainstream media.

Haag's romantic view of sports radio is a telling argument. Many fans of sports radio articulate what Haag is espousing. In fact, Haag, a feminist academic who obviously is not targeted by the niche marketers of sports radio, found that the hosts effectively patrol the boundaries of the show and cease any conversation that presents unsolvable political hatred. She surprisingly concludes that the ethic of fandom may serve as a blueprint "for what theorists had imagined as an ideal politics and public discourse of civil society" (Haag, 1996, p. 461). The ethic of fandom is one, according to Haag, in which people can speak both fervently and politely. To her own bewilderment, she admits to being hooked on sports radio while writing her dissertation, finding the shows comforting and stress reducing. Equally, at times I find sports radio helpful as an antidote to a stressful and busy career. I have developed imaginary relationships with hosts and callers that have provided a sense of belonging.

Susan Herbst (1995) also suggests that, while it would be naïve to think that talk radio allows for true participatory democracy, the genre's potential for dialogue and unstructured expression is far greater than in conventional media because of the format of the call-in programming. Sports radio presents itself as a place for callers to freely "air their take," yet it simultaneously constrains that participation within a range of in-studio control techniques. The station manager I interviewed openly admitted that the interchanges between hosts and callers is "very much orchestrated" by call screeners, engineers, and producers. "Bad callers on AM are like bad records on FM. We can't have them, so we screen them out," the station manager acknowledged. Thus, sports radio is never the truly two-way equal interaction it represents itself to be. This reality begs the question of whether sports talk radio is truly participatory or whether it gives listener/callers the feeling of "pseudo-civic participation" that "may actually thwart real participation in civic discourse, such as attending town council meetings, contacting state or federal elected officials,

discussing issues with a neighbor, and other related activities"
(McKenzie, 2000, p. 201).

The utopian view of sports radio, as a site of civic participation, has
also been challenged by both Goldberg (1998) and Mariscal (1999).
Mariscal believes that increasing corporatization of radio and the consequences of national syndication have destroyed any semblance of
the localism Haag observed. Instead of promoting locality and difference, sports radio is increasingly a friend of neoliberal ideology.[3]
Goldberg's polemic, even bolder, argues that sports talk, rather than
promoting free expression, promotes uniformity and threatens democracy in insidious ways: "Sports talk radio provides a covert political stage for those who think of themselves as nonpolitical or as politically disenfranchised. Like Limbaugh, though more discreetly, sports
talk radio enables White men to express themselves as White and
male" (p. 217). Goldberg's thesis cautions us not to think too optimis-
tically about the democratic potential of mediated versions of public
engagement. Moreover, Goldberg reminds us that perhaps the popularity of sports radio lies not in its democratic appeal but in its power
to recruit men into the values and practices associated with hegemonic
masculinity.

Once again, the female producer's observations of sports radio lend
merit to the critical standpoint taken up by Goldberg. Throughout my
interview with her, she mentioned how sports radio hosts, staff, and
callers become involved in sports talk to project their superiority over
women. Rather than appreciating the civility and respectfulness of the
discourse, my interviewee felt that men used sports radio to boast
about their own knowledge and expertise:

> Guys think they know everything about sports and everything else . . . they
> love to debate . . . they don't listen to each other . . . just talking over each
> other . . . and everyone is a better coach than the coach that is currently doing
> the job . . . although many [hosts and callers] are out of shape and not athletic,
> they can live through their favorite players and prove their male superiority.

These above comments bear an uncanny resemblance to Goldberg's
observations (1998). He writes, "Men's investment in spectator sports
accordingly becomes an investment in their own projected superiority
through the superiority of the best athletes (who just happen to be
men)" (p. 218). Sports talk is multivalent and contradictory. Surely,
sports talk does offer men an opportunity to bond and connect, but
Goldberg and Beth's comments help to highlight some of its detrimental

aspects. It was a reminder that I am not outside the masculinist discourse of sports talk and fandom, even though I intermittently experience sports talk as an opportunity for connection. Her comment highlights who is included in these bonding activities and who is not. I appreciate her openness and candor and learned a great deal about her inside view of sports radio.

ROMANTIC BELIEF IN SPORTS

Scholar Nick Trujillo (1992) argues that a sports ethnographer needs to interpret the meaning of sports from multiple points of view— namely, romantic, functionalist, and critical perspectives. The romantic views sports in an idyllic and nostalgic way, suggesting that sport is timeless and transcendent, linking us historically to an ideal past. Functionalists, many of whom are sports sociologists, suggest that sports do not teach us about an idealized past but serve as a socializing vehicle that reinforces values congruent with American capitalism: individualism, teamwork, and achievement. Critics highlight the negative aspects of sports, focusing especially on how sport produces commercialization and militarization, engenders sexism and racism, and is a tool of control for the state. According to Trujillo, all three angles are needed to study the role that sport plays in society.

The sports radio staff I interviewed—with the exception of the female producer—articulated an uncritical, romantic view of sports. When asked how they got involved in sports talk, most described a passion and love for sports and how it teaches boys—ignoring girls—important lessons in life. All five of the men I interviewed stated they were athletes who had hopes of becoming professional athletes. All but one interviewee majored in broadcast journalism and had an interest in a sports media career when their hopes of a professional athletic career ended. This romantic view of sports helped my interviewees cope with the pressures of the business end of sports radio. I asked each of them, "Why do you work in sports radio in spite of ambivalences, demands, and stress of sports radio?" Their common response: "Because we love sports!" Their career path stories also had a familiar ring. For instance, Kevin Wheeler was the play-by-play announcer for University of Miami football while attending college. After graduating, he worked his way up the ladder from intern to phone screener, to producer, to fill-in air host, to full-time talk show host.

All my subjects, with the exception of Beth, viewed sports as a necessary and positive socializing agent for boys and men. For instance,
both local sports hosts (their programs air in different demographic
areas) and the station manager made similar points:

> *Local host:* I wanted to get into sports radio because I *love* sports. It is such an
> important aspect to life, particularly baseball which I see as a living soap opera
> . . . That's the beauty of sports, it brings back an uncomplicated time . . . it
> brings back memories—memories such as the smell of a hot dog at a ball
> game. Sport connects generation to generation, grandfather, father, and son.

> *Local host:* Sports talk is how I live. I still have conversations with my dad on
> a weekly basis where we talk about the Cubs or the Bears . . . It's how we com-
> municate . . . like tons of other fathers and sons . . . that's how we bond . . .
> without sports, what we would talk about? And that's what I do on the air.
> Men need sports.

> *Station manager:* Sports talk is so mainstream now. Even President Bush was
> an owner of a baseball team [Texas Rangers]. I imagine he and Rumsfeld,
> Cheney, and Ashcroft sit around the Oval Office, drinking a cold one, and talk
> about sports when they are not talking about Iraq.

These comments fit with Sabo and Jansen's (1998) suggestion that
talking about and watching sports "is one of the few transgenerational experiences that men and boys, fathers and sons, still
share in the post-Fordist economy" (p. 205). These transgenerational
experiences are cherished moments for many boys and men in U.S.
culture, including myself. For many years and still today, I call my
dad every week to talk about the Tigers, Lions, and Redwings. Our
conversations about past sports memories help connect us to our
hometown, Detroit, even though neither he nor I have lived there for
years. Perhaps sports talk, particularly in a post-Fordist economy—a
financial system that produces fragmentation, instability, a collapsing
of time and space, and frequent geographic moving for career reasons—helps restore a bond to a local time and place. Farred (2000)
suggests that in "a postmodern moment in which people frequently
move away from their original community and continue to move
between locations, sports affinity provides one of the few contemporary forms of geographical and psychic permanence" (p. 102) My
identification with the Detroit Tigers, for example, stands as a marker
of my "home," a place that exists very powerfully in my memory; team
affiliation enables me (and other fans) to always identify with one
place—Detroit—as opposed to the reality of displacement.

In addition to the romantic notion that sports talk serves as an important primer for masculine socialization in contemporary times, the local and national hosts I interviewed believe that sports talk is an equalizing force in our society. Alan Eisenstock (2001), who interviewed several hosts and producers of sports talk radio for his book, *Sports Talk: A Journey Inside the World of Sports Talk Radio*, asked Lee Hammer (program director of the popular national sports radio show, *The Ticket*) a similar question that I asked in my ethnography: "Why do we listen to sports talk radio?" Hammer replied: "Sports is our common denominator. You can be a blue-collar worker and you can talk sports on equal level with the chairman of a Fortune 500 company. You can't talk business that way, or world politics that way, you can't talk about anything else that way. That's why we listen" (p. 215).

Even some critical sports scholars, such as Sabo and Jansen (2000) and Grant Farred (2000), agree that sports talk can momentarily break down barriers of race, ethnicity, age, and class. In his article analyzing sports talk discourse, Farred argues that "sport facilitates the transient construction of alliances across racial, class, and even ethnic lines: White suburbanites, inner-city Latino and African American men can all support the New York Knicks or the Los Angeles Dodgers" (p. 103).[4]

Again, from my personal experience, I partially concur with Hammer and Farred's comments. Throughout my adolescence and adulthood, sports fandom has provided a way in which I have been able to connect with other men across class and racial lines. At various sites— sporting events, sports bars, parties, and work—I have engaged in many spirited conversations with men of color about sports. I can recall many conversations with African American men from Detroit in which race may have been rendered temporarily insignificant; it was our identity as Tigers or Pistons fans that took primacy.

The idealistic view expressed by sports talk hosts and producers does have its merits; boys and men (and increasing numbers of girls and women) learn values such as discipline, skill, courage, loyalty, teamwork, and fairness. Sport also has a rich history that can generate deeply evocative memories and connect one to a permanent place. Lastly, sports talk can temporarily displace one's primary racial, cultural, or ethnic identity. While appreciating this perspective, I also want to critique it. For instance, the romantic outlook on sports suggests that sports exists outside power, ignoring the reality that sports talk "is freighted with political import" (Farred, 2000, p. 99).

Sure, sports talk can temporarily break down barriers of class and race, but it just as easily recreates racism with its constant reiteration of a white, color-blind insistence that race does not matter. Further, how do sports and mediated sports represent gender and nation? The romantic version of sports talk, expressed by my male interviewees, overlooks the actuality of sport as a key site in the reproduction of male hegemony and nationalism. The station manager's remarks about George W. Bush's love of sports makes evident that relationship between imperialism, militarism, and sports. Sport has figured both as a metaphor for war and a representation of national identity. Cashmore (2000) discusses how organized sport has been used by various nations as a training aid for the physical demands of war and as a tool of cultural imperialism. Jansen and Sabo (1994) draw upon a number of examples from the Gulf War of 1990 to demonstrate how American football language became interchangeable for military operations. For instance, as bombers returned from dropping bombs on Baghdad, they described the attack as a "big football game." They also note that General Norman Schwartzkopf talked about the ground war intervention as a "Hail Mary play in football." Likewise, George W. Bush has used sports metaphors in describing the recent Iraq war. The image of Bush and his administration discussing war plans in Iraq and then shifting to sports talk demonstrates this hegemonic and exchangeable link.

In terms of masculinity, several interviewees acknowledged that the practices and values of traditional manhood expressed in sports radio were to be celebrated, not disparaged. One of the local hosts I interviewed suggested that sports radio is "a guy's escape . . . a big, fun fraternity party." Many implied that since men "naturally" enjoy sports, sports radio is just logically responding to an innate need. Mike Thompson, a sports radio host, articulated something similar in the book, *Sports Talk:*

> We're men and we need "guy talk." Everything that comes up in our lives is intermingled with sports. We are re-creating what happens when a bunch of guys sit down in a barroom. It's real. We don't sit down and say, "Okay, we are now going to have the sports part of our conversation, now we're going to have the pussy part of our conversation." It's all blended. (Eisenstock, 2001, p. 42)

The idea that sports radio is just a big fraternity party has been echoed in other mediated formats—beer ads, *Maxim* magazine, and *The Man*

Show, to name a few. According to *Sacramento Bee* popular culture writer, J. Freedom du Lac (2002), the values associated with college fraternity life—beer, sex, gadgets, hedonism, and objectifying women—resonate with young men. This demographic—eighteen-to-thirty-four-year-old white men—is, according to du Lac, "a group that TV executives, magazine publishers and radio programmers are chasing in their never ending pursuit of ad dollars" (p. E7). In the article, du Lac interviews Robert Thompson, president of the Popular Culture Association, who says:

> The kind of piggish attitude that prevailed pre-Alan Alda, pre-Anita Hill, pre-PC sensitive guy days—it's all back again, and it's in a new era when you can do much more. After the period of the sensitive male who could cry at a movie and who knows not to use "he" all the time, it was inevitable we'd get a backlash. And it's really hitting the mainstream now. (du Lac, 2002, p. E7)

While acknowledging the bonding and intimacy that can occur among men in fraternities and other similar locations, this overly optimistic attitude obscures the reality that some fraternities are dangerous places for women and the likelihood of rape there is high (Boswell and Spade, 2001). In their study of fraternity parties, Boswell and Spade (2001) found that women were at a higher risk of rape in frat houses where the members promoted alcohol misuse and the degradation of women. The analogy of a fraternity party suggests that sports radio can serve as a site of misogyny, even though it is cloaked as an escapist, celebratory expression of mediated masculinity.

Nevertheless, all five men I interviewed were fairly dismissive of the accusation that sports talk is sexist. Most suggest that sports radio is just "innocent entertainment"; that the masculinist discourses expressed on sports radio programs were not to be taken seriously but, rather, ironically. For instance, Kevin Wheeler said that "guys have always sought out 'guy stuff' and they will continue to do so. The difference lately is that irreverence is more acceptable, which is probably the result of backlash against the political correctness that has bothered so many in recent years."

Wheeler's observations seem to concur with the idea, espoused by Douglas, that talk radio is a mediated "backlash" response to feminism. Wheeler's reference to "political correctness" is a thinly disguised code word for feminism. Feminism or political correctness attempts, according to some men, to deny men's natural desire for "guy stuff." Wheeler's comments can be viewed through the prism of what Beck (1997) calls "constructed certitude." Plainly stated, constructed certitude is a means of shoring up a clear and unified sense of identity by casting out or ignoring ambiguity or ambivalence. In men's lifestyle magazines, this is most clearly realized as a form of gender certitude predicated on a form of biological essentialism (Jackson, Stevenson, and Brooks, 2001). Constructed certitude provides a sense of stability amid men's current insecurities and anxieties. The construction of certitude offers a magical resolution to questions of identity, eradicating doubt and uncertainty in a society that is perceived as increasingly fragile and ambiguous. For Beck, the project of reinforcing clear and dichotomous categories of gender is less about shoring up patriarchal power and more about a psychic response to the fragmentation of traditional institutions such as heterosexual marriage and the family. Using Beck's framework, Wheeler (and other sports talk hosts and production staff) respond to charges of sexism in this way: sports radio is not sexist but merely echoing and honoring their listeners' natural masculinity and desire for "guy stuff." This outlook implies that masculinity (and male consumer desire) is fixed and ahistorical. Yet, the process of naturalizing heterosexual masculinity hides the reality that sports talk radio is not merely reflecting a "natural" manhood but helping to construct it.

What's more, the genre's pervasive use of irony and irreverence—a feature of constructed certitude—is an additional shield in response to sexist allegations. The content of sports radio is not sexist, according to many I interviewed; it is just harmless fun and satirical. The people who criticize it are failing to grasp the genre's fun and innocent intentions. In other words, a person who claims that sports radio is sexist is missing the point of the joke ("wink, wink") and taking the genre too seriously. Thus, irony operates as a strategy to dismiss feminist critiques of sports radio.

However, the one woman (a producer and fill-in host at a local station) I interviewed had a different perspective on the issue of gender and sports radio. While appreciating the multiple readings, ironic pleasures, and ambivalences within sports talk radio, Beth did not agree that it just was "harmless fun":

> It's very sexist . . . sports talk radio and sports in general . . . Sometimes the jokes on the shows go too far and offend women . . . Also, I've been told, after I was breaking down the X's and O's of the game, that it is not very sexy . . . I've also been told that I made other guys look like "pussies". . . If a guy [host] says something stupid, nobody pays attention to it . . . if a female says something stupid, it's because she's a "female". . . A lot of athletes don't take you

seriously and they think that you are just there [in the locker room] just to check them out . . . but because they don't take you seriously, they sometimes talk with their guard down and say things they wouldn't say to other media.

Beth's critical comments are in stark contrast to the romantic views espoused by her male peers. She challenges the idealistic notion of sport's transcendent and timeless elements. In addition, her comments challenge the notion of sports radio as simply consumptive fun. The sexist tenor in sports radio and, by implication in the sporting world in general, can have real consequences, even if the sexist talk is fashioned by irony. Her remarks make transparent some of the practices of sexism in the sporting world, implying that the institutional norms within the mediated sport support a patriarchal worldview; objectivity and sports knowledge are naturalized as "male" qualities. Furthermore, her remarks bluntly describe the challenges and difficulties faced by women working in sports journalism—in particular, not being taken seriously for sports knowledge and being hassled by male athletes.[5] Judith Cramer (1994) interviewed several sports journalists who describe similar experiences, especially incidents of sexual harassment in the locker room. Cramer was impressed with her informants' courage and fortitude. Despite their experiences of marginalization, they have all fought hard to earn the respect of athletes and the sport media industry. Likewise, Beth is not a passive victim; she uses her female subject position as a resource to get athletes "to say things they wouldn't to other media."[6] Her experiences, like many other women journalists who fight for acceptance within the sports world, speak to her resourcefulness and covert resistance.

HANGING OUT AT THE STATION

Most media industries, including sports talk radio, *naturalize* their process of production. Yet, for a media scholar, when something is naturalized, made to seem normal and ordinary, we most need to study it. One of my roles in this book is to question the norms and values of the sports talk radio industry and subject the workings of the media to serious analysis (referred to in the academy as denaturalization or deconstruction). One way to *denaturalize* the sports talk radio production process is to spend actual time observing the inside of a sports radio studio during programming. The process of studying people's behavior in the workplace is called *participant observation*.

As stated earlier, asking people about their work in the sports radio industry gave us certain insights, yet it has its limitations. Observing the behavior of people in that industry can be an ideal research method. Scholars use this method to examine the decision-making processes at work, the norms and values of media workers, and how the ideology in their work gets translated into media content. Participant observation has an advantage over interviews in that I am observing firsthand, not just relying on my interviewee's report of their behavior.

To chronicle firsthand the inside workings of sports talk radio programming, I called up the program director of a local sports station and asked if I could observe a show. To my surprise, he was very open to my attending a show, and invited me to the Monday 7–9 P.M. broadcast on November 17, 2003. This show is emceed by a popular host, whom I will call Lee (names are changed to preserve confidentiality). I have listened to Lee's show several times, enjoying some of the show's playfulness, his knowledge of sports, and his friendly "on-air" demeanor. The following describes my experience that night.

The station is located in a long, one-storied, nondescript building on a busy road. After parking my car, I walk around the building looking for a place to enter. Just then, a white man in his early thirties, stocky, wearing a UCLA sweatshirt, comes outside and walks to his car. Remembering that the station's update announcer, Jim, is a UCLA graduate, I call out to him, and he looks up at me. I am grateful that it is the update announcer.

"Jim, I'm the student doing a book on sports radio and George (the program director) said I can sit in on Lee's show," I nervously utter.

"No problem, come on back, the show's about to start."

We thread our way through a maze of dark hallways. I think to myself that the glamour of sports radio is in my head—the studio is truly unremarkable. Eventually we arrive at the studio, which has a long, metal table with three computers and a larger, flat-screen television that is airing the Monday night National Football League's game between the San Francisco 49ers and the Pittsburgh Steelers. Adjacent to the studio, on the other side of the glass, is the control room, small, dark, and empty except for a lone engineer, a tall, young bald man with several tattoos.

"Have a seat." Jim motions me to a chair across from him. He offers me a headset that I put on while the program director, a forty-something thin, tall white man wearing a plaid shirt, joins us. Just before the show goes to air, the host Lee arrives, a thin man in his early forties, wearing a sweatsuit. Looking wiped out, he quickly acknowledges me, shakes my hand, and sits in his chair as he slips on a headset as loud heavy-metal music begins to play. I am familiar with the song, as it begins every show that Lee hosts.

"Man, I am tired," Lee whispers to George and Jim as the guitar riff fills my headset.

George turns to me and says, "Lee and his wife just had their first baby. He's not getting much sleep."

"I can relate to that—I have a nine-year-old son. I remember those days," I say, wanting to fit in with the group.

"*Good evening, everybody, this is Lee. Welcome to the show, and we have a lot of ground to cover!*"

Lee sits five feet from me. I am struck with his resonate, lively, and animated voice. Just a few moments ago, he was looking tired and withdrawn. Lee does not begin the show with a discussion about sports, but rather the inauguration of Arnold Schwarzenegger, who was sworn in as governor nine hours prior. "Hey, listeners, what do you think about our new governor? Let's take our first call. Hi, Sam."

"Hey, Lee, I think it's great. Davis was a wimp. We needed somebody strong in there. Ah-nold [mimicking an Austrian accent] will shake things up!"

"So you like it, Sam. I don't know, Sam. I was pissed he didn't invite me to the inauguration. I thought he was going to be the 'people's governor' and yet only a few celebrities and privileged people could attend. It was roped off! And he has all those Hummers! They're such gas guzzlers! I don't know. I think I am going Green Party. I have changed since my daughter has been born." Lee takes more calls about the governor, all pro-Arnold. Eventually, Lee takes a call from the first woman caller, Susan, who wants to change the subject.

"Hey, Lee, what are you doing to me? You're killing me! I thought this was a sports station. Get off the politics. I need my sports, Lee!"

"I am sorry, Susan, but my show is more wide-ranging. I am a renaissance man. I am probably the only sports talk host who reads other sections than just the sports page. But what would you like to talk about?"

Susan wants to talk about Terrell Owens, the 49ers star receiver who is having a good game against Pittsburgh tonight. After several calls about the 49ers and the Monday night football game, the show goes to a break for commercials. At the break, I ask Lee about talking about politics.

"My attitude on sports radio is my callers like to talk about sports, but other things, too. I'd say I talk 70 percent sports, and 30 percent what's going on in the world. I don't think I have ever done a show just on sports. It's like hanging out in a bar with your dude friends. You talk about sports but other things, too."

"I also noticed that it was a woman caller who got the show back on sports. You tend to have a lot of women callers—more than other shows?"

"Yes, that's true. I get pressure to just do 'guy talk' by my manager, but I have to remember that there are women sports fans, like Susan, who listen. I know women don't bring in the advertising bucks, but I encourage them anyway."

"So you don't do the typical guy talk. You mean saying sexist things, or cutting off callers?"

"Exactly, I don't tear down my callers. I respect them. I don't cut them off. I know that tearing up a caller can increase ratings, but that's not important to me."

We then discuss ratings. Lee seems to be unflustered by them. In his local market and time slot, he maintains, his show is number one. Lee keeps his voice modulated during the breaks. He is very friendly, courteous, and answers all my questions politely. But I cannot help noticing that he is more connected to his callers; despite his engineer's occasional urging to "cut off" a caller and go to a station break, Lee is patient and listens intently. He seems genuinely interested in what they say and there is a bond between them. Perhaps Haag's (1996) thesis about civility and participatory discourse rings true on Lee's show. When a caller ends the call, Lee always thanks him or her for calling and tells them to call anytime.

The show moves on. More NFL and some NBA talk and even a few hockey questions (Lee is from the Midwest). Over the headset (which only we in the studio can hear, not the radio audience), the engineer reminds Lee that he needs to plug a new sponsor, a cigar store. During one of the breaks, I ask him about the pressures to bring in advertising revenue.

"I don't worry about it too much since we are number one in our time slot. But it's a hassle. I had to plug these cigars when I don't even smoke them anymore." During the break, the engineer, who has made several "off-air" homophobic and sexist comments throughout the evening (calling the update announcer a "fag" and saying "women with mustaches are dykes"), peeks his head into the studio and tells Lee and his update announcer that he found a new porno site while surfing the Internet. "Tits," Lee replies. Then he looks at me and says, "I can't talk like this anymore. I have a daughter." I am reminded of what Lisa Guerrero, the previous update announcer of the masculinist television show, *The Best Damn Sports Show Period* and *Monday Night Football* sideline reporter, said in a newspaper article about her experience in working with her male co-peers on the fraternity-like television sports show: "They don't live in the world that's the set of the show. When they do the show, they're belching and being dudes. But they do go home to their families. They're nice married guys who pull out their chairs for their wives. They're not like the guys in the beer commercials" (du Lac, 2002, p. E7).

Guerrero's remarks, along with the performances of masculinity by the men in the studio, remind me, once again, of the contradictions, ambivalences, and insecurities of contemporary masculinity. Heterosexual men, such as Lee, have been influenced by feminism and the "New Man" and are more involved with family life, but are still tempted into a fantasy world there they can get together and perform more laddish versions of masculinity, often through irony, playfulness, and satire. Lee, wanting to reject sexist comments due to fatherhood, suggests that he may prefer male privilege, but is making some effort to face up to postfeminist realities. Like many other men who are unsure about their place in the (post)modern world, Lee is attempting to reinvent his version of manhood, including being a responsible father.

The other issue that quickly became obvious to me while hanging out in the studio was whiteness; Lee and everybody else in the studio were white males. David Shields (1999), in his study of race and the National Basketball Association, said that "it would be impossible to overstate the degree to which sports talk radio is shadowed by a homosexual panic implicit in the fact that it consists entirely of out-of-shape white men sitting around talking about black men's buff bodies" (pp. 49–50). Race, particularly whiteness, was never discussed on the show as it seems to violate the universal sports talk code of color blindness. Yet, on two occasions, callers referred to the body of basketball player Chris Webber as "chiseled, beautiful, and athletic." The fascination with (sometimes referred to as "fetishization") black male athlete's bodies is commonplace in sports, which helps to discount the structural benefits afforded white Americans

(Mercer, 1994). What's more, the viewing of these bodies, as Shields states, is quite homoerotic. Yet these homoerotic feelings must be diffused by homophobic othering (such as the homophobic comments made by the engineer). As Eve Sedgwick (1990) has noted, homosexuality is the open secret that must be invoked and then "closeted' to protect homosocial relations (such as the men bonding in a sports radio studio).

The show ended with a call from one of Lee's regular callers, Ron.

"Hey, Lee, it's Ron. I am at Chevy's (the sports radio station hosts *Monday Night Football* viewing parties every week where a host is present along with two scantily clad women, the "Sports Girls," who hand out raffle prizes to the patrons) and the Sport Girls are awesome! You should be here, Lee! The 49ers rule. Later, Lee. I'm out."

"Good night, Ron. And goodnight everybody."

The heavy-metal music I heard at the beginning of the show is once again blaring as Lee shakes my hand, signs the confidentiality research form, and exits the studio quickly and quietly. I leave and see Lee getting into his SUV and driving away. As I turn on my car radio, the heavy metal song is still playing.

When I arrive home, I take some notes and reflect on my experience in the studio and my interviews with sports talk radio personnel. Many of the people I met in the industry experience ambivalence and contradiction in their work that, in some ways, mirror the insecurity and paradox of contemporary masculinity. While they enjoy their jobs and love sports, they are pulled in different directions concurrently: trying to address listeners as a group of regular sports fans while targeting specific market niches; viewing sports in romantic ways while being aware of the politics of radio ratings; being assumed to be primarily committed to the callers, while courting advertisers with access to a large market of consumers; providing entertaining sports talk, framed as "guy talk," without antagonizing respectable advertisers and team owners or franchises; and trying to be "one of the guys" while simultaneously making an effort to be more gender sensitive and accountable.

Surely the "guy talk" on sports radio is overtly sexist. Often, irony is a response to accusations that the sports talk radio discourse is sexist. My interviews confirm this. Moreover, it has been my experience that many men navigate the current gender order through satire and irony as a defense when they are accused of sexist practices (e.g., "I was just joking!"). The ironic content is an important aspect of sports radio's commercial success and a significant part of postmodern culture. I now will turn to questions of content and text, including the issue of irony and its role in contemporary masculinity, in Part II.

PART II

READING
SPORTS TALK RADIO

4   4th Inning

THE JIM ROME SHOW
"Myspace.com" for Men

In this chapter I look more closely at sports talk radio's content in terms of the constructions of masculinity that it represents. My textual analysis forms something of a bridge between my analysis of the production of sports radio (discussed in the preceding section) and issues surrounding consumption (covered in Section III). Specifically, I conduct a textual analysis of *The Jim Rome Show*, the most popular nationally syndicated sports radio program. I have chosen *The Jim Rome Show* due to its widespread popularity and particularly because it employs a unique language each weekday on the three-hour radio program that lends itself to a fascinating textual analysis.

According to cultural studies scholar Alan McKee (2003), textual analysis is making "an educated guess at some of the most likely interpretations that might be made of that text" (p. 1). The textual analysis McKee describes is a method whereby scholars attempt to understand the likely interpretations of texts (films, television programs, radio programs, magazines, books, clothes, advertisements, etc.) made by people who consume them.

How do we as media researchers discover the likely interpretations of a text? In order to attempt to make sense of a text (e.g., a music video or a television advertisement), the first and most important aspect to remember is *context*. There is no way that we can attempt to understand how a text might be interpreted without first asking: interpreted by whom, and in what context? McKee (2003) suggests that in order to interpret a text, one should look at its genre—the codes used to communicate between producers and audiences by following particular rules of meaning. McKee suggests that the codes and rules of speech of a text produce specific meanings that help shape one's identity (e.g., race, class, and gender). In addition to studying the genre of a text, McKee suggests that the text needs to be analyzed in the wider public context in which that text is circulated—what is happening in the larger political and social environment at the time the text is produced or consumed.

This chapter specifically examines how the mainly male audience of sports talk radio might make sense of the codes, messages, and themes on *The Jim Rome Show.* I will explore these messages and codes using the framework of Dell Hymes's (1972) idea of a *speech community*—a community that shares particular rules for the conduct and interpretation of speech that help create group identity. Specifically, this chapter will examine how the communication patterns of *The Jim Rome Show* produce a certain kind of a group male community (Tremblay and Tremblay, 2001). To further this analysis, I will examine how the show's messages, especially those concerning manhood, are located within the wider social context about what it means to be male in contemporary society. Prior to my analysis, I will provide a brief biography of Jim Rome and the history of his show. The following section will furnish background about the show's significance and parameters.

JIM ROME: HIP SPORTS TALK RADIO HOST

According to sportswriter Ashley Jude Collie (2001), Jim Rome is the "hippest, most controversial, and brutally honest voice" (p. 53) in mediated sports. In addition to his nationally syndicated radio program that airs on more than 200 stations, the forty-two-year-old hosts ESPN's *Jim Rome Is Burning,* a weekly one-hour television sports talk show (and his second show on ESPN). Rome began his radio career broadcasting University of California, Santa Barbara (UCSB) basketball games. After graduating from UCSB in 1986 and serving seven nonpaying radio internships, Rome earned a local weekend job at XTRA 690 in San Diego, a powerful 77,000-watt station. The clever fashioning of a streetwise persona, his raspy voice, staccato delivery, and fiercely independent opinions separated him from the talk-radio crowd and he soon moved into hosting a primetime radio show. Eventually, his popularity earned him a television spot on ESPN2 *Talk2,* a cable show that Rome hosted in the early 1990s.

Rome's reputation of aggressive masculinity with unassailable expertise and authority was embellished in 1994 on the set of *Talk2* while interviewing NFL quarterback Jim Everett. During the interview,

Everett knocked Rome off his chair after Rome taunted Everett by calling him "Chris" (i.e., female tennis star, Chris Evert, a veiled reference to the quarterback's reputed lack of toughness). Rome's reference to Everett as "Chris" on the show was not his first. In fact, Rome used this term for Everett throughout the 1993 NFL season on his local radio show on XTRA. This hypermasculine event increased Rome's fame and reputation among some of his fans as a host who "tells it like it is," even if it means insulting someone. However, many in the media criticized Rome's lack of professionalism and predicted the end of his career (*Sports Illustrated*, April 1994). Although Rome left ESPN2 soon after the Everett incident, his radio career slowly continued to grow. The Noble Sports Network syndicated Rome's radio show in 1995 and Premiere Radio Networks acquired the rights to the show one year later. Rome was named one of the top 100 most powerful individuals in sports by *The Sporting News* (ranked #79 in 1999). He has made a cameo appearance in the Michael Jordan film, *Space Jam* (1996), appeared in a music video for Blink 182 on HBO's television show about sport agents, *Arliss*, and cut his own CD called "Welcome to the Jungle." Rome also hosted Fox Sports Net's *The Last Word*, a sports talk television program that ran from 1997 to 2002. Rome's new ESPN television program, *Jim Rome Is Burning*, premiered in May 2003. Rome's reputation as intolerant and abusive continues to this day due to his rapid-fire, masculinist-laden opinion on sports. OutSports.com—a Web site that caters to gay and lesbian sports fans—refers to him as "the commentator who makes a name for himself by saying stupid things with an obnoxious style, that for some reason, attracts many straight sports fans" (July 13, 2000, p. 5). Rome promotes himself as the "expert" who controls the conversation with his community of listeners, much in the same way that popular "hate speech" radio talk show hosts Rush Limbaugh and Howard Stern vigorously control their program.

The masculinity performed by Rome and valorized by his clones also fits with many features of 1990s "laddish masculinity" arising in soccer fandom (Jackson, Stevenson, and Brooks, 2001). While many scholars have critiqued laddism as just another clever way of making money by resurrecting the working-class lad who was hell-bent on having a good time, others still regard the trend as a genuine rebellion, a reassertion of essential masculinity. Yet no one can deny its continuing influence, especially its import to the United States. The success of magazines such as *Maxim*, which broke the two million circulation level, making it the most successful magazine launch ever, have exploited this working-class machismo and hedonism for huge earnings (Beynon, 2002).

Rome's relationship with laddish masculinity, however, is inconsistent and tenuous. Rome, as part of the Baby Boomer generation, appears to be influenced by the 1980s "new man"—a masculinity that encourages gender sensitivity and emotional connection and caring. The famous King of Smack displays his postfeminist gender sensitivity by inviting female activists such as Martha Burk (head of the National Women's Council Organization) and lawyer Gloria Allred to speak on both his radio and television programs, where he respectfully interviews them on their views about gender and sports. Often, Rome morally condemns athletes who engage in drug use and other hedonistic acts and criticizes athletes who engage in violence against women. Rome also does not tolerate homophobic comments on his program (these issues will be explored in detail later in this chapter). This denunciation of drugs, homophobia, and violence could be taken as a sign that the hegemony of laddish masculinity is not so secure and not to be celebrated. However, Rome often engages in sexist and misogynistic jokes, such as negatively referring to tennis star Martina Navratilova as "Martin" for her performance of female masculinity (Halberstam, 1998). Rome, often through irony, encourages the clones to call the program and make sexist jokes such as referring to ice skater Tonya Harding as a "skank."[1]

Hence, Jim Rome's masculinity is contradictory and many-sided, paralleling the ambivalence that many men experience in these highly mediated and fragmented times. In addition, Rome's recent persona appears to be compatible with some of the features of contemporary "metrosexuality"—a new version of straight masculinity that combines narcissistic consumerism with a particular emphasis on style, poise, and fashion. On his program *Rome Is Burning,* Rome borrows from metrosexual masculinity by being well groomed and wearing slick sports coats with designer T-shirts or an occasional dress shirt underneath. Casual, smart, hip gear adorns the King of Smack; the cast of *Queer Eye for the Straight Guy* might rate Rome as A+ in style and coolness.[2] His ads on his radio program include Rome's promotion of tools to trim unnecessary body hair and other products for men to spruce up their appearance. In addition to a fashionable style, Rome has created his own definition of cool. His appeal is vast, as he slides into the urban lingo of a rapper and then flips the script and engages in intellectual banter with an established sportswriter and columnist like John Feinstein. Rome's multidimensional and highly performative masculinity is one to emulate—he is hip, knowledgeable on a wide range of subjects, tolerant of diversity, and easily relates to men from various backgrounds, which adds to his reputation and fame.

It is important to note that despite the variety of venues in which he appears, it is the radio show's format that contributes most to Rome's notoriety and popularity. *The Jim Rome Show*, also known as "the Jungle," has a large male audience, operates a Web site (www.jim rome.com), sells merchandise such as T-shirts and hats, performs "tour stops" nationwide (live shows featuring Jim Rome and "famous" athletes), and serves as a site for predominantly male debate about sport and occasionally social and political issues. His syndicated three-hour radio show (Monday through Friday) airs on over 200 radio stations nationwide to over 2.5 million faithful listeners (according to www.jimrome.com). Rome conducts interviews with athletes and coaches, and he regularly fields calls from listeners. In his interviews with athletes he is nonthreatening, curious, and dialogical, always treating the guest with respect (with some lapses, like his past interview with Jim Everett).

Unlike the star treatment his guests receive, however, callers engage in little respectful dialogue with Rome. Loyal callers, whom he calls "clones," phone in with their opinion (referred to as a "take") on what's happening in the world of sports. As opposed to other talk radio programs in which some dialogical interaction occurs between the caller and hosts, Rome and his callers do not engage in a back-and-forth interchange. The callers' comments are highly performative, full of insider language, and monological.[3] Rome silently listens to the call and only comments once the caller is finished with his or her monologue or Rome disconnects the call. Rarely, if ever, does a caller disagree with Rome.[4] "Huge" calls are those that Rome considers good "smack" speech—his term for sports talk that is gloatful, uninhibited, unbridled, and mimics mastery of Rome's homegrown "Jungle" speech. Callers who are run appear to accept Rome's form of hazing; Rome simply states the rules and expects the clones to accept and play by them.

According to Rome, only the strong survive in this three-hour dose of smack and irreverence. Rome's mixture of contemporary American slang is so detailed that it requires over 200 separate glossary entries on his home page. Some terms include: "fishhack," a poor print journalist;

"props," giving credit; "monkey," a radio affiliate program director; "blowin' up," becoming popular; and "jungle karma" ("positive jungle karma" occurs when an athlete appears on the show, while "negative jungle karma" is laid on someone declining an appearance). Rome's in-group language and his unique interaction (or lack there of) make his radio show distinctive. His "survival of the fittest" format is responsible for the show's reputation as the sports version of hate speech radio (Hodgson, 1999). Through his style, manner of speech, behavior, and show structure, Rome declares himself "King of the Jungle"—and only the fittest clones survive.

SPEECH CODES AND THEMES: LEARNING HOW TO SURVIVE IN THE JUNGLE

My textual analysis borrows from Dell Hymes's (1972) organizing scheme to study the speech codes and communication styles produced on *The Jim Rome Show.* To accomplish this, I taped the program from February 9 to February 20, 2004 (ten three-hour broadcasts). Parts of the show were then transcribed for more intensive analysis when I looked for particular patterns of speech acts or events that seem to be directly governed by rules or norms by this speech community. What follows is an analysis of some of the main schemes of the speech code for the listeners of *The Jim Rome Show.* I will suggest that the show provides a place for a speech community of male bonding and camaraderie. Consistent with speech code theory, I will present how the pattern of communication enacted on the program reinforces group cohesion. While much of the camaraderie and group cohesion created on the Rome show reinforces some features of hegemonic masculinity, my analysis will note some of the resistances and fissures of traditional specifications of manhood.

The overt content of *The Jim Rome Show* is sports, which, I have argued, plays a fundamental role in the construction of American male identity (Messner, 1992). In fact, "sports have become one of the last bastions of traditional male ideas of success, of male power and superiority over—and separation from—the feminization of society" (Messner, 1987, p. 196). Still, sports and sports talk do more than that: "A fundamental motivational factor behind many young males' sports strivings is a need for connection, closeness with others" (Messner, 1987, p. 198). Sport offers a "psychological safe place" (Chodorow, in Messner, 1987, p. 198) where men can connect with others but still maintain a separation within the boundaries that sport establishes. The rule-bound, competitive, hierarchical world of sport offers boys an attractive means of establishing an emotionally distant (and thus "safe") relationship with others (Messner, 2002).

At the same time, the show often goes beyond sports and integrates discussion from popular culture, music, and politics. During my two-week taping, roughly 80 percent of the show focused exclusively on sports while 20 percent focused on popular culture and sociopolitical issues. Here are a few "takes" from the clones that discussed and focused on popular culture or politics:

- Atkins had heart problems? Big surprise! Eating sausage, eggs, and cheese all day would cause heart disease (email to show, February 11, 2003).
- You mean that Princess Leah, Darth Vader, and C3PO are not real? Signed, the Weapons of Mass Destruction! (email to show, February 11, 2003).
- I have been called an "Uncle Tom" for holding other African Americans accountable when they are late for work (February 11, 2003).
- Hell yeah, Hank [Hillary] Clinton is tough. She kicked my ass. Signed, "Vince Foster" (email to show, February 13, 2003).[5]
- The only way to make a "whale" of an issue more than the Star Jones proposal would have to be what I thought was the star bangle banner sung by Christina "Skank"uilera [Christina Aguilera]. Now let me tell you, I really don't know her, so I am not making any kind of judgment, but I'm so sick and tired of people warbling the National Anthem (February 16, 2003).

So, while the manifest function of *The Jim Rome Show* "is to talk about sports, its latent content function works to construct traditional masculinity as the show and its host collectively provide a clear and consistent image of the masculine role; a guide for becoming a man, a rulebook for appropriate male behavior, in short, a manual on masculinity" (Tremblay and Tremblay, 2001, p. 278).

THE "JUNGLE": A SITE FOR THE PERFORMANCE OF MASCULINITY

Historically, male bonding has occurred in public places, such as a town square, a tavern, or a locker room (Wenner, 1998b). Oldenburg

(1989) suggests that in modern society, people require a "third place," a term he uses to describe public places that host "regular, voluntary, informal, and happily anticipated gatherings of individuals" (p. 16) beyond the home (first place) and work (second place). Third places— restaurants, coffee shops, community centers, shopping centers, and bars—function as informal public spaces that relieve stress and fashion identity in response to the alienation and nature of work in an industrial economy. Oldenburg claims that third spaces play a vital role by: (1) creating community: (2) providing a place for exchanging information; (3) helping to develop intimate friendships and; (4) serving as political and civic forums. Oldenburg suggests that third places are mainly sex-segregated sites for men to engage in male camaraderie.

While not addressing gender and power issues, Oldenburg believes that third places allow men to meet crucial needs of intimacy, community, and friendships. He notes that in the "absence of an informal public life, people's expectations toward work and family life have escalated beyond the capacity of those institutions to meet them" (p. 22). Hence, he argues, married couples face more stress as the relationship is unable to meet all the needs and expectations that are best met by an extensive social network. Oldenburg also discusses how ruling elites have often been aware of the political potential inherent in third places and have actively worked to discourage them. Sweden's ruler, for example, "banned the drinking of coffee in the eighteenth century. Officialdom was convinced that the coffeehouses were 'dens of subversions' where malcontents planned revolts" (Oldenburg, p. 23).

Rome's smack talk occurs in a metaphorical third place—"the Jungle"—a mediated site for the clones to bond and articulate their manhood. The Jungle "seems to create boundaries within which members of the community enact a particular speaking style characteristic of the community and to which its residents [the clones] should conform" (Tremblay and Tremblay, 2001, p. 13). The Jungle enacts a set of speech rules and ways of talking for the members of this airwave community. According to Philipsen, "speech is viewed as a resource for signaling one's similarity to his friends and for confirming his loyalty to them and their shared values" (1976, p. 199). Speech is also viewed as an "instrument of sociability with one's fellows, as a medium for asserting communal ties and loyalty to a group, and serves— by its use or disuse, or by a particular manner of its use—to signal that one knows one's place in the world" (p. 25).

Tremblay and Tremblay (2001) argue that the notion of Rome's show being in "the Jungle" seems to recreate a primitive place, a retreat for men to reclaim a prefeminist masculinity. They write:

> The Jungle is connected to this image of wilderness, frontier, or testing ground for masculinity. It provides a preferred viewpoint from which to view the world of the texts and constructs a position the spectator must occupy in order to participate in the pleasures and meanings of the text. Jim Rome constructs The Jungle for his audience, a place where manhood can be tested by participating in a take, or where camaraderie can be obtained, where like-minded men meet in the "wilderness" of the airwaves. (p. 279)

Tremblay and Tremblay draw upon Kimmel's (1996) idea that men are continually searching for an authentic experience and deep meaning, leading them back to the idea of a frontier or nature, a place that is needed for men in response to industrialization and changing expectations for men.

Therefore, "the Jungle" exemplifies Oldenburg's (1989) notion of a third place, an informal public place apart from work and home, where like-minded men can meet via the airwaves through an imagined community and express traditional male values. Sports talk radio may serve as this sex-segregated place for men to bond; such places are disappearing from the current (post)modern landscape:

> The reorganization of work in the Industrial Revolution has helped move friendship from shops and nearby pubs to homes. Zoning segregates big plants and offices from residential neighborhoods. Co-workers commute from many different neighborhoods and no longer come home together after work . . . Just as urbanization once fostered public community among men, suburbanization now draws them away from public community. (Wellman, 1992, p. 78)

MALE RITE OF PASSAGE ON *THE JIM ROME SHOW*

Masculinity scholars have suggested that men often engage in rituals to initiate men into manhood (Clatterbaugh, 1997; Kimmel, 1996; Messner, 1997; Newton, 2005). In Rome's Jungle, there is an initiation rite that is rule-bound and contains the possibility of failure that clones can overcome to become full-fledged members of the group. This initiation rite is the "take"—the effort by a caller to survive in the Jungle by expressing his opinion in a manner that is deemed acceptable by the expert, Jim Rome. If the caller's take "does not suck" and Rome judges it as acceptable, he says, "Rack him," and the call is considered for the "Huge Call of the Day." If Rome judges the call to be unacceptable, the caller is cut off; failure equals a "run." During my analysis, fifteen of the ninety-nine calls (15 percent) were run.[6] These code rules help the listeners understand whether someone belongs in the speech community and help members of the community make sense of others. Code rules "explain why certain takes are accepted and others are run. If a participant in the speech community has a rule violation during his take, then Jim Rome, who acts as a regulator of the speech community, negatively sanctions him" (Tremblay and Tremblay, 2001, p. 280).

The following "Huge Calls of the Day" exemplify how a listener can succeed in following these code rules in the presence of the regulator, Jim Rome:

I hope you had a good weekend, Romey. I hope you took some time to carefully consider everything that Grover Cleveland did for this country! I have two takes, about Star Jones; I have nothing but well wishes. Actually, I don't know who she is or why she's in the public limelight, I didn't know that participants of competitive eating had become so popular. But at least I'm glad she's hooking up with some guy who has bank because it's going to take some money to reinforce the structures of your home. And as far as the A-Rod trade, this is my view of it: teams are complaining of the dominance of big market teams, but other big teams such as the Dodgers or Braves had an opportunity to pick him up. To say it's not fair is ridiculous, particularly for Boston fans because you had six weeks to pick up A-Rod and work on this. There are two problems with baseball right now. First of all, there is a luxury tax, but the owners are not accountable to where they put their money. And secondly, there is no cap in the collective bargaining agreement. If they fix that, then they may have an opportunity to fix things. Until that happens, then you're going to have teams doing what they can within the rules to make sure they put a good product on the field—just as the Yankees are doing. (February 17, 2004)

I was listening to show today, and you made some great points about how this A-Rod deal has been good for baseball. But I think if baseball really wants to prosper from this thing, they need to market it in the best way possible. And, that's right, reality TV! You, Romey, said yourself that it's booming right now. And if people like Ron Jeremy [porno star] can get reality TV, then why can't FOX and MLB hook us up! I want to see A-Rod and Derek Jeter arguing in the locker room about who has the nicest suits. I want to see Steinbrenner's mood when he's gets angry and throws caviar on the walls when his Yankees go on a losing streak. And Jim, you better not hope that there were bowlers listening to you yesterday when you were blasting them! The fact of the matter is they think bowling is a great cardiovascular workout. And, after all, being the true athletes they are, they have to have some way to work off all those beers and nacho cheese Twinkies they knock off all day at the alley. (February 18, 2004)

Hey, Rome, glad to see you survived that monsoon out there. Forgive me for stating the obvious, but it sure seems that the Los Angeles Snakers [referring to the Lakers] should give Philip Garcia [Jackson] and give that guy an extension. Just pay the guy and keep him stoned. If they want to keep "Shrek" O'Neal, LA needs to pass the peace pipe and lock up Phil. Page two—baseball is as corrupt as Anna Nicole Smith and a legal team in a South Florida retirement home. Steroid scandals, players corking their bats and lying about their age, gay porn . . . Willie in KC, it must be my birthday, did you cross the line or did you snort it? Are you freaking kidding me? You're sorry for dropping anti-Semitic bombs that make Reggie White look like Billy Graham. Once a bigot, always a bigot. Are you through with Hitler's message now and turning toward Charles Manson? Did you accidentally lose your brain while picking your nose or scratching your *ass!* Have you been inspired by recent developments with Pete Rose and come up as the a-hole that you really are? Just go back to your shack and your Dick Vermeil shrine. Your act is tired. (February 19, 2004)

"Huge Calls of the Day" are recognized for their creativity, their use of smack and "gloss," their efficiency of delivery, their irony and humor, and their unique take on an issue. In contrast, callers are run for talking too long (Rome, February 13, 2004), having the radio on too long (Rome, February 11, 2004), or for saying something too offensive (such as homophobic or racist comments). A clone must succeed in the area of content and delivery to triumph in the Jungle.

This emphasis on following the speech rules of the Jungle is consistent with Tannen's (1990) observation that "for males, conversation is the way you negotiate your status in the group and keep people from pushing you around; you use talk to preserve your independence" (p. 3). Men often use communication techniques and speech patterns to prove themselves and demonstrate their knowledge and expertise. Nelson (1994) suggests that sports talk is one way that men prove their masculinity: "When they talk sports, they usually report-talk: they offer information, competing to establish who is most informed. It's a verbal one-upmanship, an oral contest. This competitive conversation simultaneously establishes both hierarchy and unity: we are men talking about men's interests" (pp. 108–109).

A "take" on *The Jim Rome Show*, hence, serves the function of an initiation ritual, which, if successfully passed, allows the clone to become a full-fledged member of the Jungle. However, the risk that the clone takes leaves him open for judgment by Rome and his listeners since being "on display is one's competitiveness . . . and at risk is one's masculinity" (Wenner, 1998b, p. 315). If one overcomes the risk and success, one is fully initiated into the "camaraderie of the group and accepted as a man among men" (Tremblay and Tremblay, 2001, p. 282). In fact, a regular caller, such as "JT the Brick," has gone on to star in his own talk radio show largely as a result of his winning Rome's first "Smackoff," a contest to judge the best caller.

IN-GROUP HUMOR ON *THE JIM ROME SHOW*

The enactments of bonding require a clone to be a regular listener of the show. A new listener can be confused by the insider language and might need to listen for several weeks to comprehend the speech rules and codes. With time, as a regular listener, one can participate in the in-group humor of the Jungle community. Peter Lyman (2001) suggests that in-group humor is a primary feature of men's relationships; "that the male bond is built upon a joking relationship that negotiates the tension men feel about their relationships with each other, and with women" (p. 145).

During the two weeks under review, one in-group joke was a constant reference to Michael Jackson ("Jacko"):

It has been almost a week and no one from the freakiest family alive, the Jacksons, has done anything to generate any news. Well, Michael, the king of freaks took care of that. According to the *New York Times,* Jackson owes Bank of America $70 million, and the money is due on Tuesday. And he can't pay. According to the report, Jacko is almost busted, and has used up almost all of his credit reserves. A financial advisor to the nutcase says that they are negotiating to have the loan extended. Banks have been reluctant to do business with Jacko since he has been accused of child molestation. That's weird. I wonder why people don't want to have financial connections to an accused pedophile. Some people are so picky. Just because he could go to the hole, not make any money and not be able to repay the loan is no reason to shut the guy out. Oh, that's right, it is. If Jacko is unable to make good on the $70 million, Bank of America would claim ownership of the Jackson music catalogue, valued between $75 million and $90 million. The plastic one is currently living in a $70,000 month rented house in Beverly Hills. Well, that makes sense. He has Disneyland about 100 miles up the road, but let's rent a house for $70,000. Maybe his secret kid room with the adult alarm is being renovated and he needed to move out for a while. Who are this guy's financial advisors, MC Hammer and Mike Tyson? Reports also say that Jackson spends $2 million per month on things like cars and shopping sprees. Good. Very nice. He also has another loan for $200 million that is heavily leveraged. Good work, Mike. Maybe you should try making another record that sells instead of going on shopping sprees and going to court. I wouldn't worry too much. I am sure that Jermaine and Tito can cover you on the several hundred million you owe. Sure they can. (Rome speaking on February 13, 2004)

References to Michael Jackson have been commonplace on Rome's show over the past three years. Because of the mutual knowledge in the Jungle, the humor does not need to be explained. Just referring to "Wacko" conjures up past references and regular "jungle dwellers" can share in the in-group humor and guffaw. Hence, in-group humor gives regular listeners a sense of community based on mutually shared background and common knowledge. The incessant focus on pathologizing Michael Jackson appears to function in maintaining group solidarity among Rome and his clones. As Meyer (1997) writes, "Humor's power in communication lies in sociability, as people share in communicating similar perceptions of the normal and abnormal" (p. 191). Ridiculing Jackson, in this sense, helps to construct the clones as "normal."

Michael Messner (2002) pointed out that men in groups define and solidify their boundaries through misogynistic and homophobic speech. Underneath this bonding experience are homoerotic and other ambivalent feelings that must be warded off and neutralized through joking, yelling, cursing, and demonizing anybody who does not conform to normative masculinity (such as Michael Jackson). Messner, in describing this homosocial enactment, writes:

> Boys and men learn to associate the group's sexual aggression paradoxically— as an exciting and pleasurable erotic bond that holds the group together and as an ever-present threat of demasculinization, humiliation, ostracism, and even violence that may be perpetrated against a boy or man who fails to conform with the dominant group's values and practices. (2002, p. 60)

Male group joking interactions are an important force in the making and remaking of hegemonic masculinity. Messner's argument (men bonding through sexist and homophobic speech) was evident during the two weeks of taping/transcribing the Rome show. In addition to many ridiculing comments directed at Michael Jackson, Rome and his clones pathologized Hillary Clinton by referring to her as "Hank" and feminized hockey player Jamir Jager.

Rome, however, ran callers who made overtly sexist comments (referring to television celebrity Star Jones as "fat") or racist comments (one caller was cut off for blaming Mexican immigrants for the loss of jobs in the United States). So, while there is a long history of male bonding through sexist and racist jokes, Rome on occasion disturbs this traditional male practice. In Rome's Jungle, members must maintain certain decorum based on a postcivil, postfeminist rights discourse that precludes racist or sexist comments, even in the form of joking (see the next chapter for further discussion of race and gender).

THE CONTRADICTIONS OF MASCULINITY

A close reading of *The Jim Rome Show* indicates that the show produces a sense of community among its listeners: mainly young educated middle-class men who have access to radio, email, and faxes during the working day. In this mediated space, a shared sense of community and a set of speech rules are created that provide a third place (not home or work) for men to connect and express their masculinity. Tremblay and Tremblay (2001) argue that *The Jim Rome Show* produces

> a speech community that appears to have morphed the traditional identity of masculinity from that of a Muscular Christian of the Industrial Age to a glib narcissist of the Information Age. This "new man" seeks to be capable and competent in Rome's radio Jungle to cope with the anxiety-producing challenges of the emerging millennium. In this constructed "place," men bond by sharing a playful speech community that has become a substitution for the real physical experience formerly acquired in the tangible arenas—the wilderness, the playing, and battlefields—for testing manhood and achieving masculinity. (p. 287)

While I agree with Tremblay and Tremblay's assertion that much of the bonding that occurs on the Rome show, with its metaphor of the jungle, reinforces notions of patriarchal manhood, the radio program may be more than simply a site for proving traditional manhood. Rather, *The Jim Rome Show* is a "third place" where men can express closeness and intimacy in a psychological "safe" place—a place where men can remake masculinity in these economically uncertain times.

Another site that has been described as a third place for men to express masculinity has been mythopoetic men's groups. Michael Schwalbe (1996), like many pro-feminist male sociologists, believed that mythopoetic men's groups (who also focus on images of wilderness and warriors) were monolithic sites of sexism. Yet, as he discovered through his fieldwork, mythopoetic groups provided a forum for men to explore and express more of the emotions that make them human, helping them see how emotions connect them to each other and the need for community and ritual.

Similarly, *The Jim Rome Show* may be a contradictory place for men: not just a site, but also a feminist backlash. Surely much of the camaraderie produced on the program fits with notions of hegemonic masculinity, but the manhood performed on *The Jim Rome Show* is not one-dimensional. The show's popularity reveals men's anxiety about finding their place in the modern world, and then seeking a "third place" to connect and even earn the respect of other men. Furthermore, the irony and masculinist humor of Rome's show may not necessarily hide a macho agenda; rather, they conceal the nervousness of men who might prefer a simpler gender and economic order, but are attempting to face up to modern realities anyway. Respect is earned not only through sexism or irony but by presenting oneself as open-minded and tolerant regarding issues of racism and homophobia, for example. Therefore, the Jungle community is many things, both enabling and constraining, including a mediated accountability community where men police each other in a postfeminist, post-civil rights America.

After listening to the show for three years, I believe it is now woven into the fabric of my daily life. I look forward to listening to it (and other sports radio programs) as it serves as a place for me, between work and home, to unwind and have an imaginary connection with other sports fans. Hence, it partially fulfills my need for a third place because it provides a place to listen to other people with related interests—sports fans. In fact, I have internalized much of Rome-speak, including using such words as "take" and "smack." Even though my study is finished, I continue to listen to the show. Although it may be difficult for progressives or academics to admit, most of us have "guilty pleasures" such as watching a reality program or reading *Cosmopolitan* magazine. These guilty pleasures do not necessarily always work in the service of hegemony.

5 5th Inning

RACE, GENDER, AND
SEXUALITY IN THE JUNGLE

This chapter will use a method of study that Douglas Kellner refers to as *ideological textual analysis*. Kellner (1995) defines ideology as a system of beliefs or ideas; all media texts are products of ideology. Sometimes the ideological position presented may be explicitly spelled out, as in overt political discourse. More often, the ideology is implicit and one has to read critically into the text to find ideology at work. The purpose of ideological textual analysis is to discover the hidden meaning(s) that may not be explicit at a first reading, often referred to in media studies as a *deep reading*. Often, ideological analysis is interested in representation—how identity is portrayed in the media (e.g., how masculinity is represented in the film *The Fight Club*).

This chapter will examine the ideologies of gender, nation, sex, class, and race expressed, perpetuated, reinforced, and resisted by Jim Rome and his listeners, providing examples of comments made by the host and callers, in the form of excerpts from audiotapes and transcripts of the show. I intend to show links between the topics of gender, sexuality, race, and class discussed on *The Jim Rome Show* and larger mediated discourse in general. I am especially interested in analyzing instances on the program that stand out as particularly important moments, what journalists often call "pegs"—critical events that generate a flurry of news coverage (Grindstaff, 1994). By examining these pegs through a deep reading of certain excerpts of the Rome show and placing them in their historical context, I hope to provide a forum in which to think through some of the ways that capitalism, hegemonic masculinity, sexuality, race, class, and consumption operate in contemporary U.S. culture.

GENDER: COMPETING MASCULINITIES

The Jim Rome Show, like much of the sporting media, celebrates traditional masculine qualities such as toughness, individualism, and aggression. Maureen Smith (2002), in her content analysis of the show writes, "On a daily basis, Jim Rome and his callers aim to define manhood and what 'real men' do. Real men do not play soccer and real men do not jump rope. Real men *do* play *real* sports." (p.11). Smith refers to real sports as the mainstream teams sports such as football, hockey, basketball, and baseball—sports that remain firmly entrenched within traditional male culture in the United States. Alternative sports such as windsurfing, surfing, and skateboarding receive little or no attention on Rome's show. Men who play these sports are feminized by Rome and his clones.[1]

Similarly, nonmainstream sports such as soccer, softball, and bowling, are marginalized as not "real" sports. For instance, on his February 13, 2004, show, Rome ridiculed retired baseball player Tom Candiotti for announcing a decision to pursue a career in professional bowling. Rome sardonically said, "Baseball is a sport, and bowling isn't. People care about baseball, and no one cares about bowling. Baseball players are athletes, and bowlers aren't . . . you're a loser if you bowl." Rome mocks Candiotti for being disloyal to traditional masculinity; real athletes ("real" men) do not play alternative sports, particularly sports that women can play alongside men (such as bowling and softball). In addition, bowling has long been associated with working-class culture; consequently, Rome is also engaging in classism.

Not only does Rome condemn men who do not play "real" sports, he passes judgment on men who engage in activities typically associated with women. For instance, Rome pokes fun at National Basketball Association (NBA) player Doug Christie for his devotion to his wife and their joint venture in designing purses:

Sacramento Kings forward Doug Christie needs to check himself, and in a hurry. Remember, this is the only player in NBA history capable of playing and maintaining a running dialogue with his wife at the same time. Some guys signal their wives from the floor during the game, like Jason Kidd at the free throw line. But Jason doesn't take it to the next level like Christie. I'm not talking about blowing a kiss or two, I'm talking about this sign language/hand gestures that Christie and his wife trade back and forth. He did it even while breaking backboards in game seven of the Western Conference Finals against the Lakers in 2002. Well, just when you thought it couldn't get any worse, it did . . . and in a big way. Mr. and Mrs. Christie are in business together designing purses. That's right, purses . . . *verrrrry* masculine, Doug! She has a handbag collection and he is helping her design the purses. I get that ballers like to have the latest gear and sport it wherever they go. I get that. But what dude says to his lady that he has some great ideas for purses? Doug Christie is forced to have sign-language conversations and design purses with his wife. Doug, you play in the NBA—you shouldn't be designing purses. You don't need the extra money and more important, you should be ashamed of yourself. What's next, bras? (November 14, 2003)

Gary Whannel (2002) asserts that, historically, sporting celebrities have been admired due to their "masculine individualism" (p. 157). According to Whannel, masculine individualism is "set against the female, the domestic and the familial, rooted in the 'naturalness' of aggression and the predatory instinct, which mother, wife, and the family threaten to tame and civilize" (p. 157). Doug Christie is feminized by Rome due to being "tamed" by his wife and for violating masculinity; no "real" man designs purses.

Nevertheless, as stated earlier, masculinity is always shifting and has multiple and contradictory meanings in different contexts (Kimmel, 1996). While masculine individualism continues to be celebrated, the sporting world has been influenced by feminist ideas about gender equality and violence against women. With the tabloidization of the media and the erosion of the division between the public and private (due in part to the second-wave feminist project of politicizing the private), greater prominence is given to "stories about the private lives of sport stars, and one by-product of this process has been the greater attention paid to domestic violence" (Whannel, 2002, p. 168). The increased tabloid-like media attention on sports stars' private lives has produced a new policing of masculinity with strong moral censure of sports stars who use illegal or performance enhancing drugs, show disrespect for rules, and engage in violence against women.[2]

Jim Rome has been quick to condemn athletes who engage in violence against women. On a regular basis, he will mention the latest athlete who has been arrested for domestic violence. Here's what Rome said on his program on February 23, 2002:

When did violence against women become acceptable? At what point in history did we officially remove any and all constraints against doing physical harm to those we allegedly love? In the past couple of weeks, there have been five reported incidents of domestic violence against women. And note, I said reported. I'm not as naïve to think that these were the only incidents that transpired. If Al Unser's girlfriend can recant less than twenty-four hours after being assaulted, then how many just don't pick up the phone at all and call the police? [Indy car race Unser apparently assaulted his girlfriend and stranded her on the side of the road several miles away from their home. Rome goes on to mention the five athletes who were arrested for domestic assault] . . . It's never OK to hit or punch your woman. If it seems that it's coming to that, take a walk around the block or find a new woman or both.

Rome has also strongly criticized University of Colorado football coach Gary Barnett for making comments about a previous woman kicker, Katie Hnida. Hnida had told *Sports Illustrated* (February 17, 2004) that she was raped by a male teammate while she was at the university. When Barnett was asked to comment about the rape allegation, he replied by criticizing Hnida's lack of football skills, never mentioning Hnida's claim of sexual violence. Many callers agreed with Rome's "take" and took positions against sexual violence.

Rome's public condemnation of athletes who engage in violence toward women and his awareness of domestic violence issues suggest that Rome's show is a mediated site informed by a competing set of ideas and discourses about masculinity rather than a single set of values promoted by patriarchy. While dominant constructions of masculinity are exceptionally present on the show, pro-feminist ideals and values are occasionally represented. The focus on domestic violence on *The Jim Rome Show* may serve as a setting for discussions about gender and violence in a place where such issues are historically not mentioned.

Yet it is important to note that Rome's moral censure of violence is contained within the mediated spectacle of the individual and the trivial; there is a lack of analysis of the broader perspective. Whannel (2002), in his analysis of moral discourse and sporting stars, praises media's recent focus on domestic violence but critiques its lack of critical analysis:

While rightly condemnatory, little of the media coverage of male violence offers much in the way of contextual insight. Issues to do with the construction of masculinity as powerful and invulnerable, the structure of patriarchal power, the concept of women as objects and as property, and the translation of human relations into commodity relations did not surface in popular discourse. (p. 172)

Likewise, while Rome's highlighting of male violence is noteworthy, his commentary lacks any insight into how patriarchal masculinity (a masculinity often celebrated on his program) creates a context for sexual assault. Rather, Rome falls into a familiar "protective" masculine discourse regarding sexual assault rather than a critical stance. Without a broader, systemic critique of patriarchal violence, I am somewhat cynical of sport talk radio's individual stories of the wrongdoings of some athletes stimulating real social change.

GENDER: WOMEN IN THE JUNGLE

Wenner (1998b, p. 303) has commented that one of the promises of sport is its place as a "naturalized refuge from women." Since Title IX, women's participation in sports has grown exponentially. This growth has challenged sports as a male preserve. In addition, the proliferation of sports media channels has provided a venue and audience for a wide range of sporting activities. However, the relative portrayal of female athletes has not improved; sports media contains female athletes by simply failing to represent them. When female athletes are represented, they are typically trivialized or objectified.

The Jim Rome Show is no different; men's sports events are promoted and described as if they have special historical importance, whereas women's sports events are typically presented in lighter, less serious, and sarcastic ways, if depicted at all. Men's events are unmarked by references to gender, whereas women's sports are always marked as "women's" events. For example, Rome refers to "women's soccer" in trivializing and sardonic ways. This is his monologue on September 12, 2003 concerning to the upcoming "Women's World Cup" soccer tournament:

> I'm a stand-up guy, and will admit for the record that I was mistaken about the sport of women's soccer. Clearly I was wrong and everybody else was right. Women's soccer is the next big thing and the women's win in the World Cup has changed the world we live in. I know this to be true because the Women's World Cup is about to start again and "only" 300,000 tickets are still available. So if there are only 300,000 available seats left, then obviously people are breaking their necks to see this once-in-a-lifetime event. So, you see, I was wrong. Any event that has sold less than half of their total tickets is obviously the center of attention in a sports sense. I mean, name one Super Bowl or World Series that sold fewer than half of their tickets. Name one! Exactly. That's because the Super Bowl and World Series aren't as important and world-changing as women's soccer. Seriously, why are people still trying to

pretend that this is something other than some little tournament? Have your players rip off their shirts in premeditated celebratory expression or whatever it is you want to do this time around. But don't try to tell me that this is an "event." It's not.

This excerpt is a classic example of trivializing women's sports in relation to the so-called importance of men's athletic events (Super Bowl, World Series). When women's athletic events are depicted on *The Jim Rome Show*, they are usually belittled, which reinforces the male preserve in sports.

Rome also frequently belittles the WNBA, stating that "the quality of basketball just isn't that good" (*The Jim Rome Show*, October 4, 2003). Even though Rome claims that "it's not a gender or sex thing, it's a basketball thing," it is definitely about gender; in particular, it is about reaffirming male superiority. Vande Berg and Projansky (2003), in their analysis of televised coverage of the WNBA, suggests that female basketball players are depicted as "exiled others," using "male standards to describe female player's athletic performances" (p. 37). In addition to bashing some of the WNBA players for not being appropriately feminine, the fan base is often mocked due to its large lesbian contingent. Fran Harris, an ex-WNBA player, has a chapter in her autobiographical account of competing in the league titled, "Why Sports Talk Guys Can't Stop Dissin the WNBA." In that chapter, she writes about sports talk radio hosts referring to the league's fans "as a bunch of lesbians":

> I've had it with tired, homophobic male sports talk radio hosts dissing the WNBA . . . Sports talk radio has produced some of the most bigoted voices in our society today. Guys who never hesitate to bash women athletes. Guys who hide behind their fears of being homosexual. Guys who can't stand that the WNBA is here to stay. I don't have a problem with these guys not liking the league. But don't stoop to insulting the fans. (Harris, 2001, p. 131)

Even though Harris did not mention his name, it was fairly clear that she was referring, in part, to Jim Rome.

There are, however, unique outcomes to Rome's trivialization of women's sports, including his 2002/2003 coverage of female athletes and men's professional golf. Martha Burk, president of the National Council of Women's Organizations, mounted a second-wave feminist campaign to include women at a private male golf club, Augusta National, the host of the prestigious Master's Tournament. Many writers and sportscasters are supportive of Martha Burk's position, including

Jim Rome, who stated that "women should have the right not to be excluded on the basis of gender" (*The Jim Rome Show,* September 4, 2002). In an interview with Burk, he asked her why she is dedicating so much time to the issue. "What about more important issues for women?" Rome asked. Burk quickly replied, "Because the issue at Augusta is emblematic of the sexism still going on in the sports world and outside the sports world" (*The Jim Rome Show,* November 21, 2002). Many of Rome's listeners were swayed by Burk's argument, calling in and articulating some of the themes of second-wave feminism—equality, inclusion, and progress. Never before, in the highly masculine arena of sports talks radio, had gender and feminism (although a depoliticized version of feminism) been discussed by so many callers.

Later in 2003, the media coverage also focused on gender and sports due to Ladies Professional Golfer (LPGA) Annika Sorenstam's playing in a professional men's golf tournament, the Colonial. Sorenstam's entry into the male preserve of golf (and, by extension, the male sporting world in general) provoked many sexist remarks by male sports commentators who believed that she could not compete with men. Old but predictable comments of men's athletic and biological superiority proliferated in the sports media, most notably by golfer Vijay Singh, who vehemently chastised Sorenstam for playing a men's tournament:

> I hope she misses the cut. Why . . . because she doesn't belong out there. If I'm drawn with her, which I won't be, I won't play. What is she going to prove by playing? It's ridiculous. She is the best woman golfer in the world. And I want to emphasize the word "woman." We have our tour for men, and they have their tour. She's taking a spot from someone in the field. (retrieved from *The Jim Rome Show,* www.jimrome.com on February 29, 2004)

Rome was highly critical of Singh, referring to him as a sexist and "old-fashioned." His commentary about Sorenstam's playing in a men's tournament was generally favorable:

> Annika has earned the respect of her peers and the gallery. In truth, she hasn't played any differently than she has on the LPGA Tour. She hits her fairways and greens . . . and doesn't make a lot of putts. Or at least that was the case in the first round. The projected cut was +1. So it looks like she needed to shoot even par. She's surpassed everybody's expectations thus far, but she needed to almost duplicate her round Thursday. I didn't even expect her to shoot that opening round of 71, so I will give her that. The truth is that Annika has proven her point. She's proven that she can play with the men. She can say she was only doing it for herself, but she's made her point to more than just herself. She

didn't finish in last and played respectably. And how bad does Vijay Singh look right about now? Sorenstam has justified being there, having the sponsor's exemption and is beating some good golfers in the process. Publicity stunt or not, Annika has earned her spot and justifiably everyone's respect. Unfortunately, Annika was not able to duplicate her level of play in the second round. She didn't hit her fairways, didn't putt like she needed, and wound up with bogeys all over the place. It went south for her very quickly and you had to know this was always a possibility. But it doesn't take away from the impact she's made. She made a respectable showing in the first round and proved her point. Ultimately it wasn't enough to make the cut, but she made a good showing.

Rome was also supportive of thirteen-year-old Michele Wie, who played well in a men's golf tournament in the summer of 2004. Rome claimed that skill, not gender, should be the issue; if a woman can compete, she should not face discrimination on the basis of her gender. In summary, while Rome's discourse often trivializes women's sports and female athletes, there are considerable exceptions. These exceptions (or unique outcomes) demonstrate that the masculinity performed on the Rome show is parallel to the discourses of contemporary masculinity in the wider culture: a set of contradictory and ambivalent meanings about manhood and a competing set of discourses ranging from male dominance/patriarchy to feminism.

QUEER EYE FOR THE SPORTS GUY

In their analysis of sport media's treatment of women, critical sport scholars have revealed how stereotyped images of femininity and heterosexuality serve to reinforce heterosexual dominance (referred to as *heteronormativity*) and trivialize women's sporting endeavors (Creedon, 1994; Griffin, 1998; Messner 2002). Moreover, homophobic representations of female athletes, particularly the symbolic erasure of women who participate in sports traditionally considered a male preserve, play a vital role in perpetuating male hegemony. My analysis of sports talk radio indicates that not only does the discourse of sports radio trivialize female athletes, it portrays women who challenge traditional gender boundaries as "unnatural" and "deviant." I will provide some textual examples on Rome's show illustrating the ways that lesbian females are represented in sports radio. I will argue that sports radio, as part of the masculine sports/media complex, maintains heterosexism by emphasizing conventional standards of white, heterosexual femininity and marginalizing female athletes who subvert traditional gender and sexual roles. These representations help produce and reinforce traditional femininities and contain the perceived threat of lesbian presence in sport. In addition, sports talk radio, by making invisible the presence of lesbians in sports, helps to naturalize sex and gender differences and reinforces ideas about women's physical inferiority. I will also discuss some ways in which the masculine performance of some female athletes disrupts and subverts hegemonic masculinity and how that is addressed within sports talk radio.

Numerous authors have made the claim that the "lesbian presence" in sport is threatening because it challenges male hegemony by upsetting existing power structures based on gender and sexuality (Birrell, 1998; Griffin, 1998; Halberstam, 1998). The question is why? Feminist scholar Monique Wittig (1993) and queer scholar Judith Butler (1990) both assert that there is no such thing as a natural category of women; women are culturally produced, not born. Similarly, Wittig argues that lesbians also are socially constructed artifacts whose existence poses a direct threat to heterosexist assumptions regarding the so-called natural connection between sexuality and gender. According to Wittig, refusing to perform heterosexuality is equivalent to refusing to become a woman. This refusal has particular material consequences for lesbians that relate to men's control over women: "For a lesbian this [refusal] goes further than the refusal of the role 'woman.' It is the refusal of the economic, ideological, and political power of a man" (p. 105).

Following Wittig's argument, sport becomes a particularly troublesome area of concern because female athletes, regardless of their sexual orientation, fit the profile of lesbians: they are frequently in groups without men; they are physically active in ways that do not have to do with being sexually appealing to men; and they are engaged in activities that do not fit with traditional specifications for heterosexual motherhood (mothers, wives).

Griffin (1998) suggests that the traditional sports media complex produces fears regarding the presence of lesbians in sport: (1) that there is an overabundance of women athletes who are lesbian; and (2) that sports participation causes females to be become lesbian. By exploiting such popularly held assumptions, those who oppose women's attempts to gain equal access to sporting resources and opportunities bring forth homophobic assumptions about lesbians running rampant in sport. Griffin (1998) suggests that a particularly effective way to prevent any challenge to male hegemony is to label female athletes as lesbians. This tactic threatens to silence and marginalize all female athletes, regardless of their sexual orientation. Women are thus discouraged from participation in sports and those who do participate are constantly navigating a homophobic landscape, making behavior choices in response to an ever-present threat of censure or ridicule. In this way, traditional gender relations are reinforced, and even as female athletes engage the domain of sports, the male preserve of sport is maintained. Homophobia regulates the behavior of female athletes and discourages significant challenges to traditionally male preserves.

Sports talk radio plays an important role in reinforcing traditional standards of white, heterosexual femininity. For example, Rome rarely interviews female athletes, and, when he does, they tend to be those who meet contemporary standards of white, heterosexual femininity. Examples include Gabrielle Reece, a skilled volleyball player who is best known for her modeling career, including her appearance in *Playboy* magazine. On the rare occasion when female sporting events earn coverage on sports radio shows, inevitably the focus turns to the athletes' femininity and adherence to heterosexual beauty standards. Heterosexually attractive women athletes are appropriated by consumer capitalism (and women's leagues like the WNBA and LPGA) to promote their sport. Tisha Pinicheiro, Lisa Leslie, Sue Bird, and other WNBA players were represented in the 2003 season promotions wearing suggestive clothing, makeup, and engaging in more traditionally feminine activities, thereby constructing a less threatening, more "family-friendly" atmosphere for their games. As media scholar Pam Creedon (1998) notes, "Homosexuality doesn't sell" (p. 96).

The WNBA's marketing strategy reflects this conventional construction of female athletics, symbolically erasing lesbians, bisexuals, queers, and women performing female masculinity (until WNBA player Sheryl Swoopes 'came out' in 2005—to be discussed later). Female athleticism is further regulated by the explicit value placed on women's sport activities. Sports talk radio, in its function of advertisement and promotion of the sporting industry, assigns particular value to sporting events based at least partly on gender. Women's sporting events are rarely covered. This gap in coverage is conspicuous in a media age in which many new or previously local and "niche" sporting events have gained national coverage.

Some feminist scholars have questioned whether female participation in sports is a productive activity that empowers women. For instance, Varda Burstyn (1999) suggests that while there is value in women learning to be active, she is uncomfortable with the hypermasculine values in sports:

> U.S. culture, influenced by men's culture, is marked by an intense denigration of the feminine and its associated qualities of softness, receptivity, cooperation, and compassion. Today's erotic flesh is hard, muscled, tense, and mean. The unquestioning emulation of hypermasculinity by women does not constitute androgyny or gender neutrality, but rather the triumph of hypermasculinism. (p. 267)

Burstyn's essentialist argument implies that women who engage in sports and take up practices that are typically assigned to maleness are reproducing the gender status quo. Her argument suggests that female athletes inadvertently internalize dominant masculine norms that colonize women's imaginations.

However, Burstyn misses out on a more complicated analysis of the ambiguous joys and potential insubordinate ways that women who appropriate masculinity through sports are creating a challenge to male hegemony. Drawing on the queer scholarship of Judith Halberstam (1998) and Jose Munoz (1999), I suggest that women who participate in sports usually associated with male physicality and aggressiveness (rugby, hockey, football, and weightlifting) are not necessarily reproducing dominant masculinity but are engaged in what Munoz refers to as "disidentification." Disidentification is "a mode of dealing with dominant ideology, one that neither opts to assimilate within a structure nor strictly opposite it" (Halberstam, 1998, p. 248). Hence, sports participation in traditionally male preserves becomes a site of cultural struggle and feminist transformation by actively disidentifying with dominant forms of masculinity and producing alternative forms of masculinity. Cox, Johnson, Newitz, and Sandell (1997), in their essay, "Masculinity without Men," concur with Halberstam's argument:

> The idea that some women might want to assume certain "masculine" traits or consider themselves as "male identified" does not suggest that women are becoming like men, but rather that the relationship between gendered roles and biological sex is more fluid than we have been taught to believe . . . Neither does such a shift automatically signal a regressive step for feminism. (p. 178)

Thus, the cultivation and performance of female masculinity through alternative sports subverts male hegemony and is not an exact replica of biological maleness.

Wheatley (1994), in line with Halberstam's argument, suggests the women who engage in sporting activities outside the mainstream sports hold greater potential for resistant sport forms. She conducted an ethnographic study of women's rugby clubs and posits that women's rugby is a site where women consciously rebelled against cultural definitions of appropriate sporting activities for women. In their appropriation of the hypermasculine rituals of men's rugby, particularly the sexist and lewd lyrics of male drinking songs, female rugby athletes infiltrate even the "malest" of male preserves. Their incursion into the rugby subculture challenges rugby practices as essential male practices. In addition, women rugby players, according to Wheatley, disturb the alliance between male hegemony and homophobia.

In the world of sports radio, however, women's alternative sports are erased and female athletes who transgress heterosexual femininity by performing female masculinity are subject to ridicule. For instance, Martina Navratilova is referred to as "Martin" by Jim Rome due to her "mannishness." When Navratilova critiqued U.S. foreign policy, this was Rome's response:

> Martin Navratilova got some issues off his chest it seems. There were some things on his mind. "The most absurd part of my escape from an unjust system is that I have exchanged one system that oppresses opinion for another . . . the Republicans in the United States manipulate public opinion and sweep any controversial issues under the table. It's depressing." Martin, you're free to leave the millions upon millions of dollars you've made at our expense and bolt for the open societies of China or maybe North Korea. I apologize for the oppressive country that is the United States. You may find that Afghanistan is more to your liking. How *niiice* of you to compare the United States to former Czechoslovakia. I didn't know it was *that* oppressive here. Thank you for enlightening me. But since you truly believe that, there are planes leaving every hour. I think you have enough money in your rather large bank account to afford a flight out. And considering that we are all about the money, maybe you should leave your millions at the door when you leave. I don't see the oppression when she is entitled to live her life, make her truckloads of money and even have the opportunity to express her misguided opinion. That's the beauty of America. She's entitled and free to express her opinion, regardless of how ill-informed, ridiculous, inappropriate, ill-conceived, unbelievable and ultimately wrong it may happen to be. She's free to do that. Not only that, she's free to leave her ungodly wealth on the kitchen table, drive her luxury car to the airport and buy a one-way ticket to either the Czech Republic or some

Middle Eastern country that espouses more "liberal" ideologies . . . especially in regards to women. If Martin is looking for a place where you can express your opinion, earn a living and lead a decent life, this is not a bad place to do it. I happen to disagree with Martin. Yes, we have our problems and issues, but relatively speaking . . . Martin has been one of the bigger beneficiaries of the American Dream. (June 2, 2002)

In this excerpt, Rome refers to Navratilova as "Martin," pathologizing her transgressive (challenging the natural gender order) performance of female masculinity as "unnatural." Because Navratilova offers a critique of U.S. policy, her transgression extends from the site of sexuality and gender to all the orders of patriarchy and the nation. Her activist voice and political critique are undermined by Rome's mockery and uncritical jingoism.

Some (Messner, 2002) have argued that there has been some slight improvement in the area of homophobia and sports. For instance, Rome has interviewed three male athletes who "came out" after they retired (I will discuss this later). Rome's focus on homophobia has privileged the gay male athlete. On a few occasions, he has commented on the strong lesbian fan presence in the WNBA. For instance, on August 4, 2002, a group of New York fans self-named as "Lesbians for Liberty" staged a kiss-in at a New York Liberty women's basketball game to protest Liberty's lack of promoting and acknowledging its strong lesbian fan base. Here are Rome's comments in regard to the "kiss-in":

But when I go to any sporting event, I could do without any public displays of affection. Man and man, man and woman, woman and woman, old dude, young woman . . . whatever. It's inappropriate in that setting. It's a sporting event. We're not talking about protesting the right of lesbians to play in the WNBA or some other right that is being denied. We're talking about the public acknowledgment of that which is private in nature. I don't care if you're into necrophilia, bestiality, or any other type of sexual preference. It's not meant for public display. (August 5, 2002)

Rome expresses a common discourse in contemporary society: *the tolerance and privacy discourse.* In the age of some increased visibility of gays and lesbians and the commodification of queerness through television shows such as *Queer as Folk* and *Queer Eye for the Straight Guy,* this liberal tolerance discourse frames the issue that we are "all just people" and hence reduces sexuality to a private, individualized matter. This discourse, according to Celia Kitzinger (1997), suggests that lesbians need to conform to the tenets of heteronormativity and assimilate to the existing social structure. Queer scholar Lisa Duggan (2002) refers to this assimilationist ideology as "new homonormativity—a politics that does not contest dominant heteronormative assumptions and institutions but upholds and sustains them while promising the possibility of a demobilized gay constituency and a privatized, depoliticized gay culture anchored in domesticity and consumption" (p. 179). Kitzinger and Duggan reject such a privatized approach because it fails to consider heterosexuality and lesbianism as socially constructed political institutions. Hence, when Rome does represent lesbian sexuality, it is contained within a liberal, private framework that reinforces heterosexual privilege (heterosexuals engaging in public display of affection at sporting events) and homophobia.

Lesbian sexuality was discussed on Rome's show when WNBA star Sheryl Swoopes "came out" in the fall of 2005. This was a groundbreaking moment in team sports; Swoopes was both the first African American and first woman to "come out" in major team sports. Rome, to his credit, did support her bravery in going public with her sexuality. However, his appreciation was contained in a way that reinforced male privilege and the trivialization of women sports. This is what Rome said on his show on October 26, 2005:

> Calling life in the closet "miserable," three time Olympic gold medalist and reigning WNBA MVP Sheryl Swoopes announced to *ESPN* the magazine she's gay. She said, "My reason for coming out isn't to be some sort of hero. I'm just at a point where I'm tired of having to pretend to be somebody I'm not. I'm tired of having to hide my feelings about the person I care about. About the person I love. Male athletes of my caliber probably feel like they have a lot more to lose than gain by coming out. I don't agree with that." First of all, I applaud her decision. It couldn't have been an easy one and it certainly was a courageous one. However, she's way off base when she says male athletes of her caliber may think they have a lot more to lose but they don't. Of course they do! A male athlete with her resume, three gold medals and a league MVP award, say, for instance, Allen Iverson, Tim Duncan, Shaq, Kobe, K.G [Kevin Garnett], any of these guys, of course they have a lot more to lose. Sheryl, you're in a fringe professional sports league and anything but a household name in this country. The guys have a lot more to lose because they have a lot more at stake. They play in a bigger league, bigger profile, bigger dollars, bigger backlash, bigger everything. I'm not looking to diminish this announcement in any way. I applaud your decision, Sheryl. It's courageous, but you're wrong to say your male equivalents don't have more to lose because clearly they do.

Rome's idea that male athletes have "more to lose" assumes a hegemonic, male perspective. Rome assumes that what is ultimately at stake is loss of sponsorships and money. This reinforces the corporate male idea that money is the definitive measure of success, an idea that marginalizes other potential negative consequences such as physical violence against women (particularly lesbians). Swoopes's social location as a black lesbian places her at potentially greater risk than white male gay athletes due to the operation of racism and sexism.

Hence, my textual analysis shows that *The Jim Rome Show* reinforces homophobia and male hegemony. However, a close reading of the show reveals some contradiction and gaps to this hegemony. The following transcripts of the program exemplify times when the show partially subverts normative masculinity and homophobia. The first example relates to an editorial letter in the May 2001 issue of *Out* magazine. In that issue, editor-in-chief Brendan Lemon stated that his boyfriend was a Major League baseball player. Lemon did not give names, but hinted that the player was from an East Coast franchise. Rome and other mainstream media programs reacted quickly to the editorial. A media firestorm resulted in a rumor mill: players, fans, owners, and sports talk radio hosts swapped guesses and anxieties over the athlete's identity.

On May 18, 2001, Rome's monologue pondered the questions: What would happen if that person's identity became public? What would it mean for baseball, gays, and lesbians in sports in general, and for the man himself? Given that Lemon's boyfriend would be the first athlete in one of the "big four" major-league team sports (baseball, football, basketball, and hockey) to come out *during* his career, what effect would this have on the institution of sport? Rome decided to pose this question to one of his interview subjects that day, well-respected baseball veteran Eric Davis.

ROME: What would happen if a teammate of yours, or any baseball player, would come out of the closet and say, "I am gay"? What would the reaction be like? How badly would that go?

ERIC: I think it would go real bad. I think people would jump to form an opinion because everybody has an opinion about gays already. But I think it would be a very difficult situation because with us showering with each other . . . being around each other as men. Now, you're in the shower with a guy who's gay . . . looking at you . . . maybe making a pass. That's an uncomfortable situation. In society, they have never really accepted it. They want to come out. And if that's the case, fine,

but in sports, it would definitely raise some eyebrows . . . I don't think it should be thrown at twenty-five guys saying, "Yeah, I am gay."
[Rome changes the subject . . . there is no follow-up.]

Rome asks a pointed question of Davis, whose predictable homophobic response warrants more follow-up questions. Yet Rome shifts the subject to something less problematic, letting Davis off the hook. After Rome ends the interview, he addresses Davis's comments in another monologue:

> That's [Eric Davis] a seventeen-year respected major league ballplayer. And I think that's a representative comment of a lot of these guys . . . He is a very highly regarded guy. This is why I asked him the question. And he answered it very honestly. He would be concerned about having a gay teammate . . . For instance, when he's showering. Personally, I don't agree with the take. It's my personal opinion. However, I posed the question to see what the reaction would be. And this is what I have been saying since this story broke. This is why it would not be a good thing. This is why the editor of that magazine clearly was wrong and has never been in a lockerroom or clubhouse. That's why it hasn't happened. Eric Davis's reaction is what you would expect. Not everybody would feel that way, but a large majority would. It would make it nearly impossible for a gay player to come out.

Here, Rome is aware of the difficulties that would occur for an openly gay ballplayer. However, he articulates his opinion in the safety of his "expert" monologue, not in the presence of Eric Davis. He does not risk compromising his masculinity or his relationship with Davis by endorsing this unusually progressive stance in the presence of a famous ballplayer like Davis. But, when a listener calls immediately after the Davis interview, Rome responds differently:

JOE: I never imagined my first take would be on gays, but I had to call. Being gay, it matters to no one but gays themselves. Why don't you guys, girls, or gays . . . whatever you guys are. Just do us a favor, do yourselves a favor and keep it to yourselves. I mean . . . [Rome runs the caller with the buzzer and disconnects the call.]

ROME: I think that's a very convenient response—"It's an issue only because you make it an issue." I don't agree with that, frankly. It's an issue because they are often persecuted against, harassed, assaulted, or killed in some cases. That's why it is an issue. They are fired from jobs, ostracized. It's not only an issue because they are making it an issue. What you are saying is keep your mouth shut, keep it in the closet; you are not accepting them for whom they are and what they are. It's not an issue because they are making it an issue. It's an issue because of people saying things like, "Keep your mouth shut . . . we don't want you around . . . we don't want to know you people exist." That's why it's an issue, because of that treatment.

Again, Rome's strong stance against homophobia demonstrates a fairly complex appreciation of the injustices of homophobia and heterosexism. This position is worth mentioning, particularly in the context of a program referred to as "the Jungle" with an audience of mostly men steeped in traditional masculinity and for whom heterosexuality is the unquestioned norm. Rome's antihomophobic stance represents a fissure in hegemonic masculinity. It can potentially foster a new awareness in Rome's listeners and invite new voices into this important conversation about masculinity and sexuality, potentially spurring a rethinking of masculinity and sports. Cutting off the first-time caller due to his homophobic comment could be viewed as a productive, accountable maneuver, which is notable since straight men do not have a rich history of holding other straight men responsible for homophobic slurs.[3]

The historic May 18, 2001, radio show generated further substantive discussion on the issue of sports and heterosexual dominance in various media sites. This included a two-part show on Jim Rome's Fox television show, *The Last Word*, titled "The Gay Athlete." The show's guests included two "out" athletes: Diana Nyad and Billy Bean. The show's discussion was very rich, with the host asking fairly nuanced and enlightened questions. Since this show, Rome has interviewed other athletes who have "come out" after they left professional sports, including football players Esera Tuaolo and David Kopay. In these interviews, Rome asked perceptive questions about the prevalence of homophobia in male sports and applauds their courage in coming out. ESPN also addressed the same topic and conducted a poll that showed that a substantial number of sports fans would have no problem with a gay athlete (ESPN.com, May 31, 2001). What's more, *Advocate* magazine published an article by cultural critic Toby Miller (2001a), who argued that the media firestorm generated by Brendan Lemon's article could potentially create a moment "for unions and owners of the big four to issue a joint statement in support, to show that queers are a legitimate part of the big leagues" (p. 3).

Another significant moment occurred on the May 18, 2001, show when Rome read the "huge email of the day," usually reserved for the nastiest comments. Rome chose an email from "Mike from San Gabriel," who wrote:

> Jim,
>   Eric Davis is perhaps the quintessential baseball player/human being who has overcome tremendous odds in battling and overcoming cancer and physical challenges. He's faced and battled a disease that strikes fear into the heart, and understands that life must be taken a day at a time. Yet, despite this brush with death and the clarity in some areas that it brings, Eric's reaction to your question regarding baseball players' reactions to knowing that a teammate is gay spoke volumes, and none of it particularly heartening. Eric's fear (speaking for the average baseball player, that is) that a gay player may be checking him out in the shower is representative of the stereotypes foisted upon homosexuals in our society, and in baseball in particular. I find it a little sad and ironic that an African American player would espouse a viewpoint—fear, ignorance, and intolerance—that for much of baseball's history had kept some of the best players in history—African Americans—out of the Major Leagues. Perhaps, though, baseball may play a progressive role in our society once again. Like it did in helping to erase the "color" barrier in the 1950s, so too it may be able to play a part in fostering tolerance and acceptance in society today. I think it's going to take someone the stature of a Jackie Robinson from the gay community to help allay the fears of baseball players, and in turn our society, before progress can be made. Until then, gay baseball players will be relegated to a shadowy world of fear and intolerance once reserved for African Americans and other minorities.
>   Mike

Mike's comments caught the attention of the editor of Outsports.com, Mike Buzinski, who commented that Mike's email of the day was "well-written" and "gay-positive." In the Web site article titled "Give the Media Good Marks: Coverage of Closeted Gay Baseball Player was Positive and Non-Judgmental," Buzinski went on to write:

> Lesbian basketball fans and gay Major League baseball players have been all the rage in the sports media the past two weeks. This alone is unprecedented. The mainstream media barely acknowledges the existence of gay athletes or fans. Having the issue raised in, among others, *The New York Times, The Los Angeles Times,* Internet discussion boards and sports talk radio is all to the good. Even better is that, overall, the coverage was balanced, informative and nonhomophobic. (p. 1)

Later in the same article he refers to Jim Rome:

> The tenor of talk radio (at least when I was listening) was not as Neanderthal as one might have expected. Jim Rome, the guy who called Jim Everett "Chris" a few years ago, has been very enlightened on the gay issue, saying it's nobody's business, while at the same time acknowledging the difficulties an "out" athlete might face. (p. 2)

Rome's stance against homophobia is groundbreaking and historic in sports talk radio.

Ultimately, however, the perspective articulated by "Mike" and supported by Rome once again confines the meaning of homophobia in sports to the intolerant or ignorant behavior of individuals, and locates the responsibility for changing that behavior in gay players or black athletes, who, after all, should "understand" about discrimination. Both Mike's letter and Rome's comments also innocently presume that African Americans have achieved equality in sports and in the larger society. This presumption, common in sports talk radio discourse, is informed by what Goldberg (1998) refers to as a "feel-good color blindness of sports talk hosts" (p. 221).[4] Queer scholars have discussed how sexuality is often produced through the process of racialization (Gopinath, 1997; Munoz, 1999). By ignoring the intersection of race and sexuality, Rome saves sports from a more biting and thorough critique, one that would expose the deep, institutional sexism and racism in sports. Instead, Rome refocuses the audience on the simple metaphors of sports—bad guy bigots and heroic gay athletes— rather than the larger environment of sports and media that keeps white, heterosexual masculinity at its center, thereby systematically excluding and oppressing all "others," including women, racial minorities, and gays.

Hence, there are contradictions in, and limitations to, Rome's "progressive" stance on sexuality. His comments espouse a liberal viewpoint that regards homophobia as fearful behavior enacted by intolerant individuals.[5] Take, for example, Rome's careless dismissal of the caller who wants gays to stay in the closet. While the caller's comments certainly reflect a homophobic viewpoint, Rome locates the blame in the caller as an individual, as if the caller is one of just a few, unenlightened bigots. A closer look at Rome's own comments on the show, including homophobic references, jokes, and name calling, all point to the same fears that motivate the caller's concern. Perhaps the caller's comments are better understood as a reasonable (but repugnant) apprehension of gays and lesbians based on the widely shared perception that "out" gays and lesbians challenge heteronormativity and patriarchy. As Claudia Card (1995) points out, hatred and hostility toward homosexuals are not a pathological disorder of a few individuals. Rather, homophobia is a pervasive problem that is not isolated in its effects.

This perspective also assumes that the right, best way for gays and lesbians to live is "out." Almost all parties in this dialogue refer to coming out, including Mike, Rome, Eric Davis, and the editor of *Out* magazine. As Gopinath (1997) observes, the "coming-out narrative" assumes that people who have a same-sex desire need to reveal their sexuality and become visible and also presupposes a universal gay subject. "Coming out" is viewed by Rome as a contested privilege, a "right," and the natural and logical next step in achieving "health" and an "authentic life." This narrative is supported by many people and institutions, including the mental health industry, straight allies, and many people in the urban gay community.

The Jim Rome Show suggests that coming out signifies freedom and egalitarianism. While this stance can provide a very powerful option for persons who identify as gay or lesbian, "coming out" can also be another standard for sexual expression that people may feel obligated to meet. In addition, privileging the coming-out narrative can unwittingly work in the service of heterosexism. Coming out requires that a person claim an identity as gay or lesbian. Foucault (1980) suggests that claming a fixed identity as homosexual may be personally liberating, but unintentionally relocates heterosexuality in the privileged center. Because straights are not required to "come out" and claim a heterosexual identity, heterosexuality is assumed to be natural and normal. While Rome and his callers discuss homosexuality, heterosexuality is never interrogated or discussed and hence remains an unmarked and naturalized category.

It is important to note that Rome's interviewing of "out" athletes such as Billy Bean and David Kopay is a unique outcome in the world of hyperheterosexual sports. To allow visibility of the gay athletes cannot be taken lightly in terms of its potential ramifications, particularly in the context (2004) of George W. Bush's proposing a constitutional amendment to ban gay marriage. Yet it is equally important to ask: Which athletes are allowed to become visible? What is their social location? How is their sexuality represented? Virtually all the gay athletes who have been on *The Jim Rome Show* are white males (an exception is Esera Tuaolo, who is Samoan) who define homosexuality as an essentialist identity. Social theorist Michel Foucault (1980) contends that while visibility opens up some new political possibilities, it is also "a trap" because it creates new forms of surveillance, discipline, and limits. Sure, Bean and Kopay are given space to discuss their experiences as gay athletes, but it must be contained within a very limited, private discourse. Scholar Lisa Duggan (2002) claims that much of the recent visibility of gays and lesbians is framed within a post-Stonewall, *homonormative* ideology. According to Duggan, homonormativity is privatizing as much as heteronormativity and each lends support to the other. As much as Rome's recognition of gays in sporting world is noteworthy, it is very much restricted within a homonormative frame that reproduces traditional gender ideas. Hence, Rome's show reinforces conservative gay and lesbian identity politics. Athletes, who perform a more unconventional, nonnormative sexuality, including women, are invisible in sports radio.

Sexuality and sports was again the subject of discussion of Rome's show on August 29, 2001. On that program, Rome was interviewing heavyweight boxers Lennox Lewis and Hasim Rahman about their upcoming title fight. During the interview, a war of words broke out because Rahman questioned Lewis's heterosexuality. Lewis became quite perturbed, stating, "I am not gay! I'm 100 percent a woman's man." This verbal conflict continued later that day on an ESPN interview program. During the ESPN taping, a physical scuffle broke out between the two boxers as they pushed each other and rolled around on the ground. The following day, Rome discussed the incident and the subsequent brawl on ESPN on his program, focusing mainly on the question of whether the incident was staged to hype the fight. Rome argued that the harsh feelings between Rahman and Lewis were "genuine"; that the incident was not staged. Yet in focusing on the theatrics or authenticity of the scuffle, Rome failed to address the inappropriateness of Rahman's homophobic slur.

The host did make an attempt, however, to address some of his callers' heterosexist/homophobic comments in the wake of the incident. On the August 30, 2001, radio show, many clones called, noting that Lewis's strong reaction to Rahman's assertion proves that Lewis is gay. Hence, homophobic gossip questioning Lewis's sexuality became the spotlight of the talk. In this next excerpt, Rome criticizes both Rahman's allegations and the callers' fixations with Lewis's sexual orientation:

Personally, I don't care. It's nobody's business what that guy [Lewis] does out-side of the ring. It's nobody's business but Lennox's. I don't care. But appar-ently, he does. He says he is not. I don't care whether he is or isn't. I tell you what—HE'S NOT GOING TO STAND FOR ANYBODY SAYING HE IS. He made that pretty clear. I don't think Rahman should have said what he said. He should not have said, quote, "That was gay of you to go to court to get me to fight." But, I tried to point out to Lennox that he's not calling you a homosexual, he's saying "it was gay to go to court." Lennox didn't want to hear it. He didn't make the distinction. And yes, it is a little peculiar that he got that hot that quickly, but I don't really care.

Here again, Rome takes up a "tolerant" position by asserting that sex-ual orientation should not matter and gossip about Lewis's sexuality is improper. Yet, by stating that sexual orientation makes no difference to him, Rome is once again invoking a liberal argument that contra-dicts his previous intolerance of the same "don't ask, don't tell posi-tion" held by a caller. In addition, his comments mirror the "don't ask, don't tell policy" on gays in the military. Scholar Ladelle McWhorter (1999) critiques this personalized approach to homophobia:

When tolerant people insist that my homosexuality doesn't matter to them, they say in effect that my homosexuality is not a social or cultural phenome-non at all but rather some sort of brute quality inherent in me and totally dis-connected from them; they say in effect that my homosexuality is a kind of ob-ject that is obviously there but has nothing to do with me as a person. Thus, this "tolerance" in the final analysis amounts basically to the same stance as that taken by reductivistic homophobes. (p. 3)

To summarize: Given Rome's prominence in the sports talk scene and the makeup of his audience, his generally progressive stance on gay athletes is significant and can be utilized as a key first step to trans-form heterosexism in sports. My intent is to recognize and respect Rome's liberal posture on sexuality while illustrating the limitations of his viewpoint. As stated in this section, Rome's stance is less than revolutionary. Obviously, I am not arguing that *The Jim Rome Show* should or will ever be a revolutionary space for alternative sexualities. Since *The Jim Rome Show* is located within the highly mediated, com-mercialized world of sport, it is unrealistic to expect Rome to take a more *queer* (more radical) position on sexuality. However, my intent is to ask questions about Rome's stance on homosexuality that have not been generally raised in the popular press. In addition, I am inter-ested in exploring how dominant public ideas about sexuality prevent us from asking more radical questions in the first place. Also, I believe that *The Jim Rome Show* and other sports media texts offer opportunities to highlight how heterosexuality is naturalized. I have often used transcripts from Rome's show to emphasize particular points on sexuality and to initiate a discussion on homophobia.

THE RACE CARD

Rome's position on race is fairly broadminded; he has a history of silencing callers who make "racist bombs" and condemns athletes who make racist comments. According to Mariscal (1999), African American callers were attracted to his show beginning in the early 1990s and began participating—a first in sports talk radio history—due in part to his antiracist stance, his appropriation of black English, and his embrace of hip-hop music. Rome was also the first national host to call upon African American sportscasters to substitute for him in his absence. Mariscal (1999) describes Rome's Jungle as a "nation that carefully policed its members to maintain a certain decorum based on post-civil rights discourse that precludes racists attacks" (p. 112).[6] During the period I transcribed and analyzed the show, Rome "ran" each and every caller who made an "explicit" racist comment, strongly admonishing the caller and saying there was no place for racism on his program.

Rome also invites scholars, such as sports sociologist Harry Edwards, on his show to discuss the issue of racism in sports. The noted sports scholar was on the program on July 9, 2003, to comment on remarks that Chicago Cubs manager Dusty Buster had made regarding black and Latino players being more able to stand the summer heat. Edwards and Rome engaged in a substantive dialogue that included deconstructing biological/essentialist views about race. Many clones called to state their appreciation for Edwards's knowledge and to discuss the legacy of racism in sports. After the interview Rome said that, "whenever we have Dr. Edwards on the show, the response is always tremendous. He provides an insight and knowledge base that is invaluable."

A key area in which issues of racism and sports are discussed on sports talk radio involves the use of Native American mascots as sports symbols. While many sports hosts and fans defend the use of such mascots, stating that the imagery is not offensive and, if anything, "honors Indians," many scholars and activists have offered critical responses on such sports mascots, claiming that these symbols are colonizing and racist (King, 2004; Mechling, 1980). Strong (2004) argues that the mascot slot assigned to Native Americans is a symbolic form of cultural citizenship that offers a considerable impediment for full participatory citizenship. Rome positions himself alongside such critical responses, stating that such images as Chief Wahoo (the mascot of the Cleveland Indians baseball club) are "offensive caricatures that are just like the old minstrelsy done to blacks" (*The Jim Rome Show,* May 18, 2002). Rome went on to say that he would give more weight to "one Native American who found Chief Wahoo offensive than 1,000 white dudes who thought it was harmless." Here, Rome is commenting on issues of power difference and white privilege. While Rome's show is marketed as pure entertainment, these examples illustrate how sports talk radio can serve as a vehicle for discussing race and diversity issues. In fact, I used this segment/transcript of Rome's commentary to address racism in my diversity class.[7]

Rome's anti-racist standpoint also includes criticizing players, coaches, and sports journalists for bigoted comments. Rome was highly critical of Rush Limbaugh for remarks he made as an ESPN analyst (Limbaugh was hired by ESPN to be the "voice of the fan" on the networks pre-game NFL television show); he said that Philadelphia Eagles quarterback Donovan McNabb was "overrated" and only praised because "the media has been desirous of a [successful] black quarterback." Rome clearly found Limbaugh's comments "racist, unnecessary, and patently offensive" (*The Jim Rome Show,* October 1, 2003). Rome is quite critical of athletes who engage in "racist bombs," particularly John Rocker, who has made several infamous racist and homophobic remarks (Rocker earned a spot in Rome's "racist fraternity house"). The following extract, dated October 24, 2001, is an example of Rome's chastising a person in the sporting world, Lee Hamilton, a San Diego radio talk show host, for making racially insensitive comments.[8] I quote Rome at length here to illustrate his rhetoric:

> He says that he made some of the comments as part of his "provocative talk radio show." Let me tell you something. Calling someone a "Jap" and saying that an African American player should be "lynched" is not provocative. It's bigoted. Now, is the old guy a racist? I don't think so . . . But he is old school. And I mean that in a bad way. I mean, join the rest of us in the twenty-first century, Lee! I haven't heard that term since the end of World War II. "Jap". . .what the hell were you *THINKING?*[9] You, Mr. Hamilton, need to enroll in those bitter old guy racially insensitive reprogramming classes. Dude

needs to get some therapy to work out his bigotry. Now, is the guy malicious? Is it hateful? I don't think so. Are those comments inexcusable? Absolutely! And let me be the first to do it since I know that you [the callers] are all fighting to be the first one to get this in. Yes, he be inducted into the Jungle "racist Hall of Fame." He's in. He can take a seat next to all the others: Jimmy "the Greek," Reggie White, Marge Schott, Fuzzy Zoeller, John Rocker, Oscar de la Hoya, and Al Campanis [people in sports who are noted for their racist comments] . . . All of them . . . Listen, I know Lee Hamilton. I worked with him for three years. Is he a malicious, dangerous racist? Do I think that he has a basement full of Nazi memorabilia and white hoods? No, I don't. But is what he said irresponsible and reckless? And does he need to be held accountable? Yes. You absolutely cannot say things like that.

Rome's monologue becomes a powerful invitation to listeners to feel involved in the discussion by calling into the show and agreeing with Rome's "take." By calling and agreeing with Rome's position, the audience can feel that it is participating in an important civic discussion of race.[10] Here's Rome's first caller after his monologue:

ROME: All right, let's go to the phones. We have a car phone in Los Angeles. Eric, you are first up. Good to have you on the show. Eric, how are you?

C: What's up, pimp? How's it going?

R: Talk to me, Eric!

C: Well, first off, I have to *GET OFF* on that crap! That is way too funny! I have been listening to you and the San Diego affiliate for years. And this guy has been dropping racist bombs as long as I can remember. It's just horrible. I'm out, Romey.

R: Good job, Eric. Rack him. Listen, here's the thing. I think that Lee Hamilton says things that are irresponsible and reckless and inappropriate. I don't think that he is a vindictive, malicious bigot, in the true sense of the word. As I said. I don't think he has a hood. I think he's old school. I think he's ignorant. And old school and uninformed in an ignorant way. But does he hate and discriminate against minorities? I would say no. I know the guy a little. I worked in the same building with the guy and I never really saw any evidence of that.

The caller expresses his own distaste of racist "bombs," earning the approval of Rome. This transcript illustrates how *The Jim Rome Show* creates a speech community in which people are held accountable for racist comments; in order to be in the Jungle you must express racial tolerance and cultural competence. Yet, Rome and his fan's comments reflect a narrow, liberal-humanist version of racism.[11] This color-blind discourse, historically central to U.S. sports, is informed by the idea that race does not matter; that due to sport's integration and recognition of athletes of color, it has transcended race. The universal sports talk refrain is that "I don't care if you're purple, orange, green or yellow"—meaning that, in athletics, everybody is equal, as long as they can perform, reinforcing the idea of meritocracy and mythology of the American Dream. In the confined world of mediated sports, antiracism is about individual niceness: if a person means well, then he is not a racist. When Rome bombastically invites Hamilton to "join the rest of us in the twenty-first century," he is suggesting that bigotry is located only in a few, unenlightened individuals. In addition, Rome privatizes the problem of racism by appropriating therapeutic rhetoric (Cloud, 1997) and suggesting that Hamilton needs psychotherapy to unlearn his racism. This ideology prohibits Rome and his fans from critiquing the institutional aspects of U.S. racism or the racist practices in the sports media world.

Rome's comments also suggest that he gets to decide who is racist or not—he is the final arbiter of justice. Rome, a white straight male, does nothing to interrogate whiteness.[12] Hence, whiteness remains invisible in the discussions on his show. This arbitration is further extended to the audience through their invitation to join Rome in the moralizing: blaming Hamilton as an individual for his inexcusable comments, equating bigotry and racism, and thereby reaffirming their own "antiracism."

The critique of Rome's liberal antiracist discourse is not intended to disparage its potential constructive effect. Certainly, in the realm of hate speech radio, racist and hegemonic masculine discourses dominate, and Rome's liberal stance is a welcome addition. However, it is also important to note that Rome's somewhat admirable antiracist stance is contradictory. While condemning callers who make racist comments toward African Americans, he has on occasion, according to Jorge Mariscal (1999), engaged in racialized stereotypes about Latinos. Mariscal (1999) theorizes about Rome's discrepancies:

> The contradictions apparent in Rome's comments need not be attributed to some generalized unpredictability of the postmodern condition or even to an ideology free sphere of mass communication. Rather, Rome's inconsistence stance on racially charged topics reveals the basic slippage in liberal discourse, a situation in which the citizen/subject struggles to maintain the decorum of

post-civil rights Black–White race relations even as he slides easily from a
tepid anti racism to the (unconscious?) reproduction of deeply ingrained racist clichés. (p. 116)

The commercial medium of sports radio diminishes Rome's tepid antiracist stance—a form deeply embedded within liberal, post-civil rights
discourse and unbridled, global capitalism.

Like Mariscal, I found inconsistencies in Rome's antiracist posture,
including many racialized remarks he has directed toward African
American athletes. Cole and Andrews (1996) note how black athletes
are stereotyped in the media as deviant, oversexed, and drug-addicted,
linking them to a threatening "urban black masculinity." Rome has reinforced such racist rhetoric, including being highly critical of African
American athletes who engage in trash talking, taunting, celebrating,
and dancing, stating that such behavior is the epitome of poor sportsmanship. For instance, when San Francisco 49er football player, Terrell
Owens, ran his touchdown ball out to midfield and placed it on the
Dallas Cowboy star in an act of taunting toward his opponents, Rome
supported the NFL's decision to suspend Owens and fine him $24,000.
In fact, Rome gave Owens the nickname "Terrellenthal," associating
Owens's behavior with O. J. Simpson (O. J.'s full name is Orenthal), a
prominent black athlete accused of murdering his wife. The specter of
race reared its ugly head here as Rome racializes Terrell Owens's behavior, linking it with O. J. Simpson's criminality.[13]

Herbert Simons (2003) notes that the extraordinary media attention these behaviors (trash talking, taunting, dancing, and/or celebrating) receive seems out of proportion to their importance, since
they provide little if any competitive advantage and seem to be only
peripherally related to the actual competition. Simons argues that the
extraordinary attention these sport behaviors receive is racially moti-
vated in that black athletes are said to be largely responsible for such
acts. These behaviors are seen to be a reflection of urban African
American cultural norms, which conflict with white mainstream
norms. In summary, Simons posits that the restrictions placed on
such behaviors represent white male society's response to the threat
to white masculinity represented by black athletic superiority and by
African American athletes' assertion of the right to define the meaning of their behavior. Rather than framing these behaviors as celebratory and potentially productive ways that African American competi-
tors resist white male norms and assert their cultural identity, Rome

and the sports media world in general condemns such behavior, claiming that it insults the so-called universal norms of sportsmanship; white dominant sporting norms, such as humility and not calling attention to oneself, are privileged. Rome and many of his callers also expressed qualms that behaviors such as trash talking and taunting could lead to "out-of-control violence." By linking such behaviors to aggressive behavior, Rome evokes the unconscious racialized fears of white sports fans that African Americans are naturally prone to violence.

The prototype "threatening" black athlete is baseball superstar Barry Bonds, who is reviled by the press. He is probably the most polarizing athlete of his generation. Bonds is on the cusp of breaking baseball's most hallowed record, Henry Aaron's home run record. Bonds is despised by the majority of sports talk radio hosts, including Rome, due to his arrogant and surly manner with the press and fans and allegations that he used steroids.[14] Much of the simplistic debate on sports media centers on whether the hatred of Bonds is due to racism—white athletes (e.g., Mark McGwire and Lance Armstrong) have been accused of steroid use, but do not receive the condemnation Bonds has. According to Professor Leonard Moore,

> White America doesn't want him [Bonds] to pass Babe Ruth's [home run record] and is doing everything they can to stop him. America has not had a white hope since the retirement of (NBA star) Larry Bird, and once Bonds passes Ruth, there's nothing that will make (Ruth) unique, and they're scared. And I'm scared for Bonds. (quoted in a USA Today article by Nightengale, 2006, p. 1)

While the debate about race, steroids, and Bonds is complex (including the issue of hypermasculine bodies, capitalism, and performance), the specter of race cannot be ignored. For example, ESPN radio host Jason Seibel (filling in on the The Dan Patrick Show) said in response to steroids, "If he did it, hang him!" (Zirin, 2006). Seibel, offering images of a controversial black athlete being lynched, conjures up the worst legacies of U.S. racism. Many black athletes, including retired San Francisco Giant Willie McCovey said, "I don't think it [steroid allegations] would be this big a deal if Mark McGwire was still playing and was in the same shoes chasing that record. I don't think they would be spending all this time to dig all this dirt up on him. Racism is a thing that we have to live with that people don't even realize" (Zirin, 2006).

Bonds himself has often openly discussed the racism in baseball. Instead of taking seriously the accusation of racism, Bonds is seen as shifting the focus onto himself and playing the "race card." In sports radio, the *race card* is a very common phrase that often works to play down a serious discussion of racism. While Bonds's accountability on steroid use may be an important part of the dialogue on steroids and sports, a sole focus on the San Francisco baseball star ignores Major League baseball's responsibility in looking the other way for many years (due to promoting the allure of the home run) as well as its long history of racism.

One way that white sports fans and, by extension, the sports media negotiate the "threat" of the so-called deviant black athlete is by venerating African Americans who are positioned as "colorless" and race-transcendent. Tiger Woods, Magic Johnson, and NBA player Tim Duncan are just a few of the black athletes who are deraced. The quintessential race-transcendent African American sports person is Michael Jordan. Jackson and Andrews (1995) critically analyze Jordan's media image:

> Through the mutually reinforcing narrative strategies employed by Nike, the NBA and multitude of other corporate interests (e.g., Coca-Cola, McDonald's), Jordan was constructed as a racially neutered (hence non-threatening) black version of a white cultural model . . . Jordan's racially transcendent image was All-American . . . Jordan became a commodity-sign devoid of racial integrity which effectively ensured the subversion of racial Otherness, but which also—because of his media pervasiveness—further ensured the celebration of the NBA as a racially acceptable social and cultural space. (p. 4)

Andrews (1996) argues that Jordan is a representative of racial displacement rather than race transcendence; Jordan's racially sanitized image is shifted onto other black bodies who become racialized (Charles Barkley, Alan Iverson, Barry Bonds, and African America athletes who are portrayed as criminal and gangster-like) or the nameless urban black male who is criminalized in the mass media.

Rome has praised racially neutered athletes, including Michael Jordan's heir apparent, Kobe Bryant. Some in the media commented that Bryant was not as "hip" and "cool" as other black NBA players, such as newer stars Lebron James and Tracy McGrady. According to this logic, Bryant might not be as marketable to young fans who overidealize the black player with "street credibility"—a term to describe athletes who have an urban sensibility with hip-hop style and who express a "tough" and, at times, sexist image. On his June 25, 2003, show, Rome made these comments in response to this logic:

> Of course we've all heard the rap against Kobe. The reasons why Kobe doesn't have the supposed "street credibility" are because of his affluent upbringing, clean-cut style, attended private schools, has no tattoos, and never been arrested . . . Don't penalize Kobe because he can speak more than one language and doesn't have children out of wedlock. Don't penalize Kobe because he's never been arrested or embarrassed his team or family.

Here, Rome commends Bryant for his presentable image and his fidelity, invoking the values promoted by the Million Man March: male accountability, monogamy, faithfulness, and a benevolent, patriarchal masculinity (Carbado, 1999). While many of these values are congruent with some of the values of feminism (egalitarianism and male responsibility), they privilege heterosexuality and sexism. Rome also uses these values to invoke racialization by contrasting Bryant's respectable image to the deviant image of other African American athletes in the media, such as NBA player Allen Iverson.[15]

Rome further discounts black athletes, who are portrayed as thuggish, particularly when they speak out against racism in sports. NBA player Rasheed Wallace, like Iverson, is depicted as criminal and threatening in the mainstream media and is regularly chastised on Rome's show. While he played with the Portland Trailblazers (Rome often refers to the team as the "Trailgangsters" because several team members—all African American—were arrested for drug possession), Wallace agreed to an interview with the *Oregonian*, in which he told the reporter that the NBA was racist. Rome was quick to discount Rasheed's contentions, claiming that Wallace was playing the "race card." According to Cole and Andrews (2001), the race card is a highly conservative notion that regulates and "governs discourses around race in the USA . . . as it implies that the introduction of racial divisions is inappropriate and unfair" (p. 76). Rome, by invoking the "race card" discourses, reinforces the American myths of individualism and meritocracy while eliding a critique of the institutional racist structures of modern sports.

Carrington (2002) suggests that the sports media complex obsesses about African American athletes, allowing white sports fans to fulfill voyeuristic desires to look at black athletes. The homoerotic desire fetishizes black athletes, reducing their bodies to commodities:

The growth in the visual spectacularization of the black male from within Western media culture raises a number of issues relating to the diminution of politics and the reconfiguration of the subaltern public sphere. These undoubtedly contradictory processes are played out within the single economy where "blackness," and the black body, have become highly valued commodities whilst at the very same time within the formal economy, actual Black people struggle to survive against the material conditions of global capitalism. Black people, once literal commodities during the Atlantic slave trade, have been transformed into commodity-signs to be bought and sold throughout the globalized media market. (pp. 26–27)

Likewise, when Rome and his clones honor the few black athletes who have experienced success in the current global economy (in particular, sporting men who have been racially neutered by the process of commodification), it fosters the belief that African Americans regularly "make it." And if they do not succeed financially, it is due to their own moral deficiencies. This dominant attitude reinforces the notion that sport is the ideal or only way for black youth to "escape" the ghetto and their natural cultural immorality (Cole, 1999). Consequently, the messages on *The Jim Rome Show* reinforce the American Dream myth, masking structural economic problems and oppressive policies while concurrently pathologizing African Americans as morally inferior.

It is important to note that while the media advantage a few African American athletes, they can easily lose their privileged position and be "re-raced" when their behavior is associated with immorality. Jordan temporarily was racialized when the media publicized his gambling problems and Bryant has been "re-raced" since his arrest for felony sexual assault in July 2003. Rome expressed his disappointment at Bryant on his show on July 21:

We thought this guy was the one guy who was different. We thought that that one guy who preached ethics and morality who actually lived that life. He conned us all . . . There was not a better guy with a better image in the NBA. He got married early and seemingly stayed away from the flamboyant lifestyle of many other players. This was the guy who was supposedly the quintessential family man.

On one hand, in this excerpt Rome holds Bryant accountable for behaviors that men historically have been able to get away with—namely, infidelity. Yet Rome's diatribe is supported by conservative, heterosexist values. Bryant's transgressions could have been an opportunity for Rome and the larger sports media to examine the contradictions and struggles of black masculinity. For instance, neither Rome nor anyone in the sports media has discussed how many black men may hold tightly to white hegemonic notions of manhood as a response to racial oppression and the promise of patriarchy; that disenfranchised black males will claim their patriarchal manhood through illegal ways since socially acceptable ways are unavailable to them. Rather, Rome's heteronormative "take" bolsters on some level, perhaps unconsciously, the image of the black male rapist and white female victim. While this may describe the specific events of the Bryant case, an analysis that includes both race and gender eludes Rome and the larger sports culture, keeping the status quo intact.

ROME HAS NO CLASS

Critical sports scholarship primarily analyzes sport through the lens of race and gender; class analysis is quite limited in the field. In her book, *Where We Stand: Class Matters,* bell hooks (2000) reminds us that an exclusive focus on race and gender can mask the brutal realities of class politics. While class issues may be underanalyzed in academic discourse, it is ever-present in sports media discourse, including on Rome's show. Rome, like many other sports journalists, incessantly focuses on the various transgressions of Tonya Harding, the infamously scorned figure skater who notoriously attacked competitor Nancy Kerrigan in 1994. Harding, perhaps more than any other sporting celebrity, is singled out for the harshest treatment by Rome. For instance, on February 12, 2003, Rome referred to Harding as a "pipe-wielding, knee bashing, home porn-making, hubcap-smacking, trailer-evicting, celebrity-boxing skank." Rome often brings his clones up to date with Harding's latest brush-in with the law or appearance on *Celebrity Boxing* match, usually beginning his monologue with "Tonya Harding is back in court, back in the news, and back in the Jungle" (*The Jim Rome Show,* January 10, 2003).

Rome's reference to Harding's class position evokes references to *white trash* stereotypes. Wray and Newitz (1997) suggest that the white trash label serves to fortify for the middle and upper classes an impression of cultural and intellectual supremacy. Since overt racism is unacceptable in the Jungle, poor whites such as Harding became a safe target to direct bigoted and racist feelings. As Laura Grindstaff (2002) writes, "White trash creates a new racial other, which, like racial others of old, is linked to notions of uncivilized savagery and thus occupies the primitive side of the primitive/civilized divide" (p. 264).

Why is Tonya Harding the object of so much scorn by Rome and other reporters? It's not just that Tonya is working class. Many other skaters have come from poor backgrounds: Oksana Baiul was orphaned at an early age and grew up in the Ukraine, and Nancy Kerrigan grew up blue collar and was snubbed by the elite Boston Skating Club due to her class position. However, Tonya is different: she is not just working class. She is not embarrassed about being working class and makes no attempt to apologize for it. Whereas Kerrigan performs "her class and her gender in a recognizable way, signaling effortlessly and naturally she was already one of us" (Foote, 2003, p. 8), Harding drinks beer, drives a truck, shows up on reality programs, and dates construction workers. In her article, "Skating on Thin Ice: Why Tonya Harding Could Never be America's Ice Princess," feminist writer Elizabeth Arveda Kissling (1998) offers an enlightening perspective:

> Tonya Harding became the butt of popular jokes and an object of journalistic ridicule because she holds up a different mirror to our society. Tonya Harding's autobiography reflects images of us we'd rather not see. Harding reflects a femininity that includes aggression, competitiveness, and muscularity, without apology. As a so-called white trash poster child, she reflects a way of life marked by violence and poverty that stretches far beyond her Clackamas County, Oregon hometown. Her efforts to skate away from that life reflect a shattered ideal of a meritocracy, a real world in which it takes more than a dream and hard work to succeed—a world where everyone is skating on thin ice. (p. 2)

Since Harding does not conform to standard cultural scripts governing class and gender, she is the object of ridicule on *The Jim Rome Show*. While some diversity is tolerated in this mediated space, the excessive cultural transgressions of Harding are too much for Rome and his clones to find their more liberal, tolerant voices. In addition, her story exposes the myth of the American Dream, which may be too threatening for some clones, many of whom are working class themselves and may also be skating on thin ice in these uncertain economic times. Tonya's representation potentially threatens men—if she can break out of her working-class position, why can't they? In addition to class subversion, her aggressive, athletic skating style disrupts the traditional femininity of most figure skaters, potentially threatening taken-for-granted ideologies of female sporting bodies.

Another area in which Rome has historically evoked white trash images is in the popular sport of NASCAR racing. Rome, in the past, has referred to NASCAR as "not a real sport . . . making left turns around an oval is not a sport." He has also referred to NASCAR fans as rednecks, poor southern whites who have nothing better to do. He even referred to NASCAR as "NECK-CAR" due to its so-called red-neck fan base. Rome rarely discussed NASCAR racing on his show in any serious way and rarely interviewed the drivers. NASCAR was only a vehicle for Rome's sardonic humor.

Rome's "take" on NASCAR began shifting in 2003; suddenly NAS-CAR was viewed as a mainstream sport by Rome and sports radio in general. Rome began to seriously discuss NASCAR and has frequently interviewed NASCAR drivers such as Jeff Gordon and Dale Earnhardt Jr. Why the sudden shift to respectability? Rome has recognized the growing popularity of the sport, noting on June 12, 2003, that it is a $2 billion a year industry that grew 65 percent in the 1990s. As a tele-vision sport, it is second in viewers only to the National Football League (Swan, 1998). According to NASCAR's Web site (www.nascar .com), more than 250 sponsors are associated with NASCAR, includ-ing seventy Fortune 500 corporations. Rome, whose show is a part of the same corporate culture, appears very cognizant of the power of corporate sponsors. He even apologized for his previous opinion on NASCAR, stating that he knows many clones are NASCAR fans. NASCAR's success is an excellent illustration of how a sport's popular-ity shifts depending on the support of corporate America.

NASCAR's recent surge in mainstream acceptance may also be due in part to an important voting demographic: "NASCAR dads." While "soccer moms" were the important voting bloc in 2000, NASCAR dads—working-class, rural, white heterosexual men who used to vote Democratic but now usually vote Republican—were courted by George W. Bush and, to a lesser degree, by the Democratic presiden-tial nominee, John Kerry. In fact, George W. Bush was present at the Super Bowl of NASCAR, the Daytona 500, where he talked about his love for the sport. Bush was the Master of Ceremonies at Daytona and had the distinct privilege of declaring, "Gentleman, start your en-gines!" NASCAR dads are represented as being very conservative, as noted by National Rifle Association (NRA) Chairman Wayne La-Pierre. According to LaPierre,

NASCAR Nation is NASCAR Nation is NRA Nation. The people at NAS-CAR races are your hard-working, average tax-paying Americans that are rais-ing their families and putting their kids through school. They are patriotic. They own guns. They hunt, and they go shooting and they love the Second

Amendment, which is what we're about, also. It's where America is, to tell you the truth. If you want to find mainstream America, go to NASCAR. (quoted in Clarke, 2003, p. A1)

Rome's new respect for NASCAR might mean a new, more tolerant view about class relations since the sport's historical fan base has been working-class southern whites. I am somewhat skeptical. It seems that his (and sports radio in general) newfound respect for NASCAR is more about the power of corporate sponsorships and conservative political values. It is no coincidence that NASCAR dads, not soccer moms, are an important voting bloc in a post 9/11 America. Similar to Reagan's efforts to remasculinize America after Carter and the Iran hostage affair, Bush, through NASCAR dad imagery, is upping America's testosterone levels after the Clinton presidency and Al-Qaeda terrorism. *The Jim Rome Show*, by extension, colludes with such neoconservative ideologies.

JUNGLE NATIONALISM

This chapter would be incomplete if it did not include an analysis of *The Jim Rome Show's* link to nationalist discourses, particularly in a post 9/11 America. Much has been written by sports scholars on the connections between sport and nationalism, how states use the symbolism of sport for purposes of nation-building (Allison; 2002; Cashmore, 2000; Coakley, 2004). Rome is no exception here; he often refers to sport as reflecting and upholding "American values."

Rome's nationalistic rhetoric has significantly increased since 9/11.[16] In the days after 9/11, Rome focused on the role sports should play in response to the tragedy. Both Major League baseball (MLB) and the National Football League (NFL) canceled games for several weeks. During this period, Rome interviewed many sports reporters on when the MLB and the NFL should resume play. Rome and his interviewees came to the conclusion that the leagues should resume play as soon as possible because the games could "heal the nation." Many callers reaffirmed Rome's "take," hoping that the show could just go back to talking about sports; that such talk could mend some of the emotion and wounds.

During the weeks following 9/11, Rome also talked about how athletes were "natural heroes" in a time of national mourning. He spent an entire hour on September 25, 2001, honoring Mark Bingham, an

athlete who led the University of California, Berkeley, to two national rugby titles. Bingham, who was "out" as a gay man, was on
United Airlines Flight 93 on September 11, 2001, from Newark to
San Francisco. The plane crashed into a grassy field outside Pittsburgh, and many believed that Bingham's athletic strength helped
prevent the hijackers from reaching their intended target. Rome suggested that Bingham's training as a rugby player, a contact sport
with no time-outs and with a need to think quickly under pressure,
helped him to interrupt the hijacker's plans. Rome said that "there's
something to be said for competitive sports; his last game was not on
a grassy field. It was on a narrow 757." Neither Rome nor his callers
mentioned Bingham's sexual identity during the program; Bingham's
national hero status would have been disrupted by the knowledge
that he was gay.

As time has passed, the discourse on the Rome show continues to
have a U.S., nationalistic focus. One way nationalism has expressed
itself on the show is Rome's continual bashing of soccer. On a relent-
less basis, Rome will talk about his main mission—"I will not stop
until I get every last soccer game and soccer league run out of America." On November 25, 2003, Rome mentioned the world's most
popular soccer club, Manchester United, was touring the United
States. Some sportswriters viewed this tour as evidence that soccer
was catching on in the United States. However, Rome vehemently
disagreed: "We don't like the game over here very much. Actually, we
don't like it at all. Soccer is catching on in the U.S? Sure it is. Soccer
sucks." When sportswriters claimed that soccer's popularity was increasing because of the 1999 U.S.A. Women's soccer team's winning
the World Cup, Rome dismissed the significance of the event by stating that the only thing worth mentioning was that winning goal
scorer Brandi Chastain's sports bra was exposed when she took her
jersey off in celebration.

Rome is not alone in marginalizing soccer. Many other sports talk
hosts also trivialize soccer by claiming that it is too boring and too
low-scoring. However, scholars Markovits and Hellerman (2001)
suggest that these reasons are not why soccer is a distant also-ran behind football, basketball, baseball, and hockey. Rather, the authors
argue that when American sports culture developed in the late nine-
teenth and early twentieth centuries, nativism and nationalism were
shaping a distinctly American self-image that clashed with the non-
American sport of soccer; baseball and football crowded out the

game and reinforced the notion of American supremacy. Markovits and Hellerman, through a thoughtful and detailed analysis, suggest that hegemonic sports culture colluded with the discourse of American exceptionalism—a particular form of self-contained nationalism that reinforces the idea that the United States is uniquely superior. Rome's scathing attack on soccer, the most global of sports, is situated within this form of nationalism. By mocking soccer, Rome ridicules others nations who have a passion for the sport (he also "bashed" soccer prior to 9/11). American exceptionalism—whether the sports media's bashing soccer or the Bush administration's going it alone in the Iraq War against the wishes of the United Nations—is alive and well in post 9/11 America.[17]

Since the 2003 Iraq War, the Rome show has periodically evoked nationalistic discourses by feminizing the friendly nations that did not support the U.S. invasion of Iraq—Canada, France, and Germany. When the 2004 movie *Miracle* premiered (a movie that chronicles the 1980 U.S. Olympic hockey team's amazing victory over the heavily favored Soviet Union), Rome talked about how the movie helped to restore "American pride" in the same way that the actual victory in 1980 reinstated patriotism during a period when the United States was demoralized due to the Iran hostage affair. On February 13, 2004, Rome interviewed one of the members of the 1980 U.S. Olympic hockey team, Jack O'Callahan, about the movie. Several minutes into the interview, O'Callahan started talking about how winning the gold medal helped the United States overcome the stigma of being "too soft." Then he dropped the bomb by saying, "I've always told the Canadian guys, that if all of the guys in the United States who went on to play basketball, football, and baseball grew up playing hockey, we would have never heard of Wayne Gretzky."

Needless to say, many Canadian listeners did not respond well to O'Callahan's nationalistic comments. Wayne Gretzky is a national hero in Canada, and many Canadian clones called Rome, loudly declaring how offended they were at such distasteful remarks. Rome defended O'Callahan, stating that his comments were said in jest and should be taken ironically. When Canadians called, Rome discounted their critiques, noting that they should just "get the joke and chill out." Irony, once again, was the defense in response to claims of prejudice and mean-spiritedness. Rome encouraged U.S. callers to feed off O'Callahan's ironic remarks and rewarded callers who said such things as: "Canadians are little bitches . . . we kicked your *ass* in the

Junior hockey tournament . . . God Bless America . . . Canada, should rename themselves America-Light." The huge call of the day was a man from Toronto who said the Canadians should "lighten up and understand irony . . . when you get pissed off on such comments you prove that we Canadians suffer from an inferiority complex."

One Canadian whom Rome admires is hockey announcer Don Cherry, another national icon, noted for tough masculinity and his "tell it like it is attitude." In March 2003, Cherry received a great deal of criticism for remarks made on *Hockey Night in Canada*. On this popular show, he criticized Canada for not supporting the United States when historically the United States has always supported Canada during wartime. Rome interviewed Cherry on April 1, 2003, about his remarks, at which point Cherry said that antiwar protestors were traitors who supported Saddam Hussein. Many clones lauded Cherry's comments, with callers saying that you cannot support the troops and be against the war. Rome, however, disagreed:

> That's too close to "America, love it or leave it." You can support our troops, wishing them to come home safely after doing what they have been ordered to do, yet not agree with why the country is involved in a war. I believe you can disagree with an administration and yet, at the same time, want our troops to do their job and come home safely. Both ideas can exist. Just because you don't agree, it doesn't make you "Un-American" or a "Saddam supporter." I can't say it enough; peaceful protesting is what America is about. I may not agree with the antiwar sentiment, but I agree with their right to protest and ultimately that it isn't "Un-American" or unpatriotic.

Rome's views in regards to the Bush administration's policies are in stark contrast to the jingoistic speech on political talk radio shows such as those of Rush Limbaugh or Michael Savage. Rome has also been somewhat critical of the Bush administration's lack of finding the so-called weapons of mass destruction. To criticize Bush, even slightly, is atypical and risky in the world of talk radio, particularly since 9/11 and even more so since the infamous Janet Jackson "breast exposure" incident at the 2004 Super Bowl.[18] Since the Jackson incident, Clear Channel and the FCC have increased censorship practices, including banning Howard Stern from several stations in a controversy many claim is due to its alliance with Bush. As Cole (2004) writes, "it's a plausible theory, given that the claims Clear Channel made about Stern's offensiveness are entirely new, although Stern has been offending almost everyone for years with the blessing of his sponsors" (p. 92). The only thing that has changed is Stern's recent conversion from a Bush supporter to endorsing the Democratic candidate (John Kerry) in the 2004 presidential election. Clear Channel has also been accused of censoring or censuring entertainers for expressing views that conflicted with those of the Bush administration. Disc jockey Roxanne Walker is suing Clear Channel for allegedly firing her for disagreeing with the president's policies in Iraq (Cole, 2004). Lowry Mays, Clear Channel's founder, has been a generous and longtime supporter of the GOP and President Bush, donating tens of thousands of dollars. Hence, for Rome to condemn a Bush position is somewhat bold and insubordinate since he is an employee of Clear Channel. Once again, the discourse on the Rome show, and sports talk in general, is more complex and ambivalent than politically conservative talk radio; there is more room for different points of view among sports fans. Perhaps sports talk radio has more potential for civic dialogue than other talk radio and media forms.

HEGEMONY OR HOPE?
SPORTS TALK RADIO'S POTENTIAL

While many of the messages on *The Jim Rome Show* reinforce traditional masculine ideas, my textual analysis shows that the program is not simply a vehicle of sexism, homophobia, and racism. This chapter highlighted some of the unique outcomes of *The Jim Rome Show*, illustrating how the show is a multifaceted, ironic, and contradictory text that includes moments of progressive, socially just discourse. These contradictions and ambivalences should not be discounted and may provide some impetus for engendering new conversations about gender, sexuality, and race. Unlike other masculinist genres—such as men's magazines or *The Man Show—The Jim Rome Show* is more complex and has the potential to generate meaningful discussions about social issues.

Following this line of reasoning, I think it is important for both scholars and progressive sport fans to be attentive in noticing and promoting innovative moments in the sports media. The antihomophobic, antiracist tenor of *The Jim Rome Show* is a chance to address male hegemony and racism in sports. Rome's show does reveal the limits of liberal/tolerant ideologies, but his progressive stance on social justice may be a starting point for influencing male sport fans who subscribe to the ideals of traditional masculinity. Rome's prestige and authority with millions of male sports fans give him leverage to effect a more progressive political agenda. There is potential here for this male bonding speech community to do more than just live a life based on the values of beer, babes, and ball. There is opportunity on Rome's show for men to engage in relationship building and to reinvent masculine ideals.

In summary, while the progressive moments on *The Jim Rome Show* are tenuous and fleeting, they still may provide impetus for socially progressive dialogue. Feng (2000) suggests that although resistance within the text is short-lived, moments of audience resistance may be mobilized "into a new narrative space" (p. 48). Hence, resistance by spectators has more substantive potential than the ephemeral progressive moments in texts. Feng's thesis fits with my experience; I have had many productive and progressive conversations with my friends and even academic colleagues that were inspired by Rome's take on a particular issue. The next chapter, an audience analysis, will examine the possibility of audience resistance within the dominant ideological textual themes of sports talk radio.

PART III

THE AUDIENCE OF SPORTS TALK RADIO

6 6th Inning

IN THE JUNGLE
WITH THE "CLONES"

In previous sections, I have examined the texts and industry of sports radio. Now I turn my attention to the third major area of cultural studies analysis: audiences. In cultural studies, the term "audience" refers to the people who attend a particular play, view a film or television show, read a novel, or listen to a radio program. The audience is also used in a broader sense, referring to people who are exposed to media culture. In a sense, the term "audience" is exchangeable with "society," for it is used to refer to the many ways in which mass media/popular culture relates to the wider social world. From this perspective, all citizens of a society make up a potential audience for a media product. In this chapter (and the subsequent two chapters), I will be considering mainly the listeners of *The Jim Rome Show* to examine some of the ways they interpret and make meaning of sports talk radio texts.

There are many ways to research audiences. The first set of methods involves observing audiences, focusing on ethnography (or fieldwork). Other related methods include asking questions of audience members through individual or group interviews. Other researchers have used questionnaires to examine the ways audiences consume a media artifact.

In this chapter, my audience analysis is based on a series of semistructured interviews with fans of sports talk radio conducted at various sports bars throughout the United States. There are inherent challenges in researching radio audiences. In particular, it is difficult to access a radio audience for participation, observation, and dialogue. Sports bars served as ideal sites for my research because many of the patrons who frequent these establishments are avid listeners of *The*

Jim Rome Show and other sports radio programs. In addition, since it is a primary site for male bonding, the sports bar is an extension of the environment created in sports talk radio where similar social practices and discourses are evident (Wenner, 1998b). Given that my research is limited to a small number of participants and that the audience members I interviewed may not be representative of the North American sports radio audience, the results are not necessarily generalizable. My hope is that my findings will promote insight into what the recent dramatic growth of sports talk radio signifies in terms of men's changing identities and gender relations.[1] I am particularly interested in exploring the pleasures associated with listening to sports radio, the imagined community that is created through sports radio, and the meanings that listeners make of some of the more contradictory and progressive moments of *The Jim Rome Show*. While my analysis found little evidence of active resistance to the masculinist ethos of sports talk, I discovered some ambivalence in the way men (and some women) interpret sports talk radio, suggesting a certain uneasiness with more hegemonic forms of masculinity. This analysis yields a view of contemporary masculinity as fragmented, ambivalent, and influenced by consumer culture and neoliberal discourses.

INTERVIEWING THE "CLONES"

I started exploring the complex relationship between audiences and texts by hanging out in sports bars and interviewing listeners to *The Jim Rome Show*. These interviews took place between July, 1, 2001, and September 2, 2001, in sports bars in Sacramento, Tampa, Las Vegas, and Fresno. I conducted semistructured interviews with eighteen people who described themselves as fans of *The Jim Rome Show*. The average age of the participants was thirty-two. Ten were Caucasian, three were African American, three were Latino, and two were Asian American. Sixteen of the subjects were male and two were female, all identified as heterosexual.[2] Twelve of the men were married and the two women were single. Thirteen of the interviewees, including both women, identified as "white collar" and were working in some area of business or sales. Three subjects were truck drivers and the other two were construction workers. Twelve of my subjects had college degrees.

Conducting research as a sports fan in the highly masculinized space of a sports bar produced some interesting ethical dilemmas, including issues of power relations and gender. Free and Hughson (2003) discuss the issues of gender blindness and other research dilemmas in their analysis of recent ethnographies of male soccer subculture (Armstrong, 1998; Giulianotti, 1999). The authors critique Armstrong and Giulianotti's research as omitting issues of gender and sexuality from their accounts of research participants (soccer hooligans). Free and Hughson wonder, for instance, "Did the young men under study trade in sexist jokes? Did Giulianotti and Armstrong respond? Did they feign laughter? If not, were they ridiculed?" (p. 138).

Taking Free and Hughson's critique into account, I attempted to be self-reflexive about my privileged subject position (a white, male, heterosexual sports fan) so as not to inadvertently engage in "male collusive discourse." This privilege was evident when I was discussing my research with a male friend who identifies as gay. As he said to me, "I could never do that research; a sports bar is a dangerous place for a gay man. I would feel very unsafe there." Taking his comments into account helps me continually to reflect on my privileged status as a researcher and as a straight, middle-class, white male fan. I also prefer to ask critical questions that invite my male subjects to examine and interrogate masculinity, in an effort to situate my research in a broader social justice context.

Being a researcher and a sports fan who has frequented many sports bars has both advantages and disadvantages. The main risk is over-identifying with my research subjects and not having enough critical distance. The main advantage of being a sports fan is that it helps to facilitate nuanced understandings and forms of access impossible from other subject positions. Conducting this study as a fan inspired me to a high degree of accountability—I have included verbatim materials, edited and selected from my taped interviews, as a way of privileging the voices of my informants (all the names of my research subjects have been changed to preserve anonymity).

The sports bar is a fascinating site in which to conduct fieldwork.[3] In his assessment of the cultural space of modern and postmodern sports bars, Wenner (1998b) argues that alcohol, sports, and hegemonic masculinity operate as a "holy trinity." He distinguishes the modern sports bar, a traditionally gendered place, from the postmodern sports bar, a locale where gender relations are rearranged into a commodified hybrid. The modern sports bar, according to Wenner, is a place in which to talk to your male peers, have a drink, and watch and discuss sports—places I remember hanging out with my father and grandfather after Detroit Tigers games. Wenner describes the traditional sports bar as one "where public drinking and participation in sports serve as masculine rites of passage, their spaces and places often serve as a refuge from women" (p. 303). In contrast, the postmodern sports bar, in addition to being less sexually segregated, is "designed as an experience as opposed to a real place" (Wenner, 1998b, p. 323). He writes:

> The postmodern sports bar does not seek to stimulate the "authenticity" of a local place. Designed for out-of-towners to catch the game and for the realization that fewer and fewer people live in the places they were from, the postmodern sports bar offers "memorabilia in the generic." A wide net is cast so that there is some identity hook for everyone, no matter what their favorite team, level of fanship, or geographic past. (p. 325)

The bars I frequented were of the postmodern type that Wenner describes. Lacking the smell (I remember the local taverns in Detroit as smelling like men's locker rooms) and look of local sports bars (worn furniture, photos of local sports heroes, and virtually all men—no women—sitting at bar stools), the bars where I conducted my research were airy, bright, lively, loud, smelled clean, and were also noted for their full menu of food. Gantz and Wenner (1995) have noted that food and drink play a major role in the sexual geography of sports bars: the more food is foregrounded, the more women tend to be present as servers and patrons. As drinking replaces eating as the main activity, the layout of the bar becomes more sexually separate, with women appearing, if at all, as adjuncts to men.

The bar areas looked very similar in each city—a large rectangular perimeter that resembled a large table with four corners and a "wet area" in the middle serviced by bartenders. And, as noted by Gantz and Wenner, the sexual geography of the postmodern bars I frequented was more egalitarian. Both men and women worked as bartenders and waiters. During the time I spent observing, the majority of people sitting at the bar were men, but women also sat there without noticeable harassment. The space was a metaphor for postmodern culture in general—a constant tension between democratization and commodification.[4] In this space, I found that male hegemony was still present as in the older bar context, but in a more understated way. As Wenner (1998b) writes, "In the postmodern sports bar, male hegemony does not go away, it is merely transformed by its reframing" (p. 327).

In each bar I visited, I sat at the large bar area and began socializing with patrons, discussing sports and currents events. After some small talk, I asked if they listened to *The Jim Rome Show*. All the men I approached stated that they listened to the show. I then informed them of my research project and asked them to do an audiotaped interview about their experience of the show. All agreed enthusiastically after I assured them of confidentiality. The interviews, generally lasting thirty minutes, were enjoyable as well as substantive and informative. Questions I asked included, but were not limited to, the following: How often do you listen to *The Jim Rome Show*? What do you like most about the show? What do you like least about the show? What do you think the show means to most men who listen regularly to the program? Why is the show popular with many men? What is your view of Rome's position on various social issues? Have you ever called the program? What do you think of the "takes" of the callers?

As I reviewed each interview transcript, I made notes about its content, analyzing the responses to each question. I was particularly interested in looking for common themes, key phrases, ways of talking, and patterns of responses that occurred in my conversations with the subjects. Instead of using a positivist model of research, I regarded my audience research as provisional, partial, and situated in a particular social and historical location. As Ang (1996) states, "critical audience studies should not pretend to tell the 'truth' about 'the audience.' Its ambitions should be more modest" (p. 45).

All eighteen of my research subjects stated that they listened to the program while commuting in their car. Thus, my research confirms Oldenburg's (1989) and Tremblay and Tremblay's (2001) argument that sports radio serves as a "third space"—a space in between work and home—for many men. The blue-collar workers (truck drivers and construction workers) stated that they also listen to sports radio while working, which serves as a bonding experience. Occasionally, according to one construction worker, there is conflict between some of the workers who want to listen to music instead of sports radio. The white-collar workers stated that they listen to the program while driving alone in their car from and to work; it tends to be a solitary listening experience. Six of these fans stated that they switch stations between *The Jim Rome Show* and Rush Limbaugh, who is on the air at the same time. These listeners identified as conservative Republicans (they were all white males). Three of the businessmen I interviewed also subscribe to Jim Rome's Web site, so they are able to listen to the show via the Internet while in their office. "Rome gets me through the stress of the day," one investment banker told me. Eight of the professional workers stated that they were listening to the program less because of their work. With the advent of conference calls and cellular phones, many feel the pressure to make business calls while commuting. All eight complained about the increased work demand that, among other effects, is chipping away at their "third place" and the pleasure of listening to *The Jim Rome Show* and other sports radio programs.

When I asked the twelve married men whether they listen to *The Jim Rome Show* with their wife and family, they all laughed, stating that deciding which radio station to listen to was a regular source of conflict. One man said, "I don't even bother; the wife hates Rome and so do the kids, so I just let them listen to music." Three of the men said that since they are driving their car, they "as the man" decided which station to listen to. Hence, the radio dial, similar to the television remote, is where gender and power relations are produced (Morley, 1992). Only one man observed that he and his wife both enjoy the show because it serves as a place of connection. "My wife loves sports," the thirty-one-year-old salesman told me. "I just needed to teach her the Rome lingo and his smack-tionary so she could understand the program. Now she listens on her own!"

Although the common initial explanation for listening to *The Jim Rome Show* was "It is entertaining," the conversations thickened as we focused on issues of homosexuality, masculinity, and other social topics. My skills as a practicing psychotherapist came in handy as I invited people to open up about particular parts of their lived experience and about controversial topics. Often, people disclosed very personal stories and thanked me for an "enlightening" or "thought-provoking" experience. The interviews confirmed the notion that sports talk can provide an opportunity for men to discuss, and even raise their awareness of, gender and sexual issues that they might not otherwise have.

What's more, the conversations were not shallow, homophobic, or sexist. Although our conversations were brief, the men were often warm and open, particularly when I asked them about their relationships, family, and children. One man in Sacramento, a twenty-seven-year old, expressed a sense of emptiness because he was divorced and not currently in a relationship. In fact, he shed a few tears when talking about missing his three-year-old daughter who lived with his ex-wife in another state. I confided that I was also divorced, but had remarried, finding the emotional fulfillment I had desired. As our conversation ended, I encouraged him to "hang in there" and to continue pursuing his wish for connection and intimacy. Other men who were married spoke fondly of their wives and children, illustrating some of the features of the "new man"—sensitive, emotionally aware, respectful of women, and egalitarian in attitude. Hence, far from objectifying conversations about women that characterize male homosocial conversations, many men spoke of their desires to create and sustain intimate relationships with women. Now I will turn to some of the emergent themes that developed from my audience ethnography.

THE ENTERTAINMENT VALUE

One common element in many of the consumers' responses was that the program was entertaining. Many found Rome's wit and aggressive smack talk pleasurable and amusing. Many appreciated the host's honesty and "authenticity." For example, one man said: "I listen every day. He tells like it is. He lets it rip. He doesn't hold back. I like that! And he's entertaining! He pokes fun at people. It's funny! It reminds me of locker room humor . . . Yes, I get a kick out of his smack talk. It's pure entertainment. Like when he trashes NASCAR and the WNBA." Another man echoed similar sentiments:

> I thought it was hilarious when he called Jim Everett "Chris." That's what sticks in my head when someone says something about Rome. He's kind of like the Rush Limbaugh or Howard Stern of sports talk radio. Like he thinks he's God. But I don't mind it because he's entertaining. And it's a way for him to get the ratings and the market share. I admire that because I am a stockbroker. You need to market yourself to stand out. You need to be aggressive and controversial to be successful in today's society . . . The show makes men cocky—like the clones. I listen to it for the entertainment. And he does know his sports.

These celebratory interpretations of the show are fairly representative of the participants I interviewed. Many valorized Rome's "transnational business masculinity," a term coined by Connell (2000) that is marked by egocentrism, conditional loyalties, and no permanent commitments other than to capital accumulation. Additionally, as stated previously, many subjects listened to the program because of the pleasurable aspects of the program, finding Rome to be knowledgeable, authoritative, and comedic. The implicit assumption was that listening to Rome was an innocent pleasure. One person, when asked about the so-called harmlessness of the program, said, "If you don't like it, turn the radio dial. No one is forcing you to listen. It's just entertainment!" This is a common response to critiques of the negative effects of media culture and audience pleasure. Yet, amusement is neither innate nor harmless. Pleasure is learned and closely connected to power and knowledge (Foucault, 1980). Media scholar Douglas Kellner (1995) writes:

> We learn what to enjoy and what we should avoid. We learn when to laugh and when to cheer. A system of power and privilege thus conditions our pleasures so that we seek certain socially sanctioned pleasures and avoid others. Some people learn to laugh at racist jokes and others learn to feel pleasure at the brutal use of violence. (p. 39)

The media industry, therefore, often mobilizes pleasure around conservative ideologies that have oppressive effects on women, homosexuals, and people of color. The ideologies of hegemonic masculinity, assembled in the form of pleasure and humor, are what many of my participants found most enjoyable about *The Jim Rome Show*. This includes Rome's aggressive, masculinist, "expert" speech that ridicules others. Thus, many of the pleasurable aspects of the program may encourage certain male listeners to identify with the features of traditional masculinity.

HOMOSOCIALITY

As stated earlier, sports, and its media partners, expanded to create a homosocial institution that served to counteract men's fear of feminization in modern culture. The interviews I conducted appear to confirm this elucidation. When asked why *The Jim Rome Show*, and other radio sports talk programs are so popular among heterosexual men, many men told me that they feel anxious and uncertain due to the changes in men's work and women's increasing presence in the public sphere. Several participants believed that sports talk provides a safe haven in which men can bond and reaffirm their essential masculinity. Here's an example:

> It's [*The Jim Rome Show*] a male bonding thing, a locker room for guys in the radio. You can't do it at work, everything's PC now! So the Rome show is a last refuge for men to bond and be men. Look, even here, in a sports bar, there are women. There is no place for men to just be with men anymore, even in a sports bar. Now, the radio, you're in your car and it's just you and Rome, and

it's the audience that you can't see. I listen in the car and can let that maleness come out. I know its offensive sometimes to gays and women . . . you know . . . when men bond . . . but men need that! . . . Romey's show gives me the opportunity to talk to other guy friends about something we share in common. And my dad listens to Romey also. So my dad and I bond also.

These comments are telling about the mixed effects of sports talk. On the one hand, sports talk radio allows men to express a "covert intimacy"[5] (Messner, 1992) and shared meaning about a common subject matter. This bonding can bring forth genuine moments of closeness and should not be pathologized or seen as completely negative. Sports talk, as stated earlier, does create a space in which men from various classes, races, ethnicities, and ages can relate to each other. Nevertheless, as the interviewee clearly states, this discursive site is "offensive sometimes to gays and women." The interviewee's remark that women have now penetrated the traditionally sexually segregated space of the sports bar is also revealing. His comments concur with Tremblay and Tremblay's (2001) argument that radio has "taken over the bar's traditional male bonding function" (p. 277) since the geography of bars is now open to women.

One of the forces that most shapes traditional masculine homosociality (Wenner, 1998b) is men's need to strive and compete for prestige and approval within their peer groups. This striving provides the basis for affiliation and camaraderie. Many people I interviewed stated that the ultimate compliment would be for Jim Rome to approve of their "take." To have your call "racked" by the leading sports media personality would be a real honor in Rome's world. However, this mindset also means that if one gets on the air and Rome disapproves of the call, it could signify a failed masculinity. To illustrate this point, here is what one man said:

> I never have called. I thought about calling but I would hate to get run [Rome's disconnecting the call]. Man that would hurt! I sometimes think, "Man, I could give a good take . . . but if I call and 'suck'. . .you know . . . get run, start stuttering . . . man that would be embarrassing. [section deleted] I think people call because it's Jim Rome. He's the man! He's the pimp in the box! Man, if you get racked and are the caller of the day, you're the man!

Interestingly enough, the only person I interviewed in this part of my study who actually called the show was a woman. She shares her experience of being on the air with Jim Rome: "I actually called. My voice was heard. He was cool. He didn't bag me. I didn't speak the clone language. I am me! I called to state that violence should stay in hockey—hockey is not hockey without violence—I feel very strong about that." When I asked her a follow-up question about what was "cool" about Rome, she said:

> He's cool to women. He's not sexist. He respected my call. I know sexist guys. I have lived with them. My ex-husband was one. He was very violent and mean. I lived in a bad environment. I left him and moved to Vegas. I no longer will be around men who are offensive or demeaning to women. I don't feel that from Rome. I don't see him as demeaning to women. If he were a sexist, I wouldn't listen to his show.

Joan's comments could be seen as firmly grounded in postfeminist discourse that perpetuates hegemonic masculinity.[6] Her comments advocate continued violence in hockey and she does not experience Rome as sexist despite Rome's repeated misogynistic references on the show. Joan presents herself as "one of the guys" and replicates a patriarchal view of the show and its contents. Her comments also reveal the notion that even though the masculine behavior of the Rome clones is that of traditional white heterosexual masculinity, those outside the group (older men, men of color, gay men and women) can gain entry into the Jungle if they are familiar with the speech codes and can modify their talk so as to diminish differences, real or imagined. Hence, while women were historically excluded from such conversations that sustain male solidarity, women can now be included if they accommodate their behavior to fit that of the group. If women call Rome, "they must play by the men's rules. Otherwise they undermine the purpose of sports-talk: the worship of man and his accomplishments." (Nelson, 1994, p. 122)

THE AUDIENCE DOES SOCIAL ISSUES

Another common theme that I identified from my transcriptions was appreciation that Rome and his show trafficked in larger social issues. The overwhelming majority of interviewees respected and agreed with Rome's opinions or "takes" on the issues of gender, race, and sexuality. For many, the show was an important forum in which to discuss and reflect on wider political matters. The following response from a twenty-four-year-old white male is an example of a "preferred" reading of Rome's liberal humanist, "don't ask, don't tell" position on homophobia: "Romey is like a sports sociologist with humor. He's entertaining. He's really into the gay issue. He's an advocate for gay rights. I respect him for it because he speaks his mind. Personally, I don't care what gays do. But I don't think gays in team sports won't work because so many athletes are macho and homophobic." All eighteen people I interviewed respected, albeit somewhat ambivalently, Rome's "tolerant" position on the issue of homosexuality and sports. In fact, several said they had not thought of the issue of homophobia and sports prior to Rome's addressing it on his program. Many respected Rome for this stance and said they would now support an athlete who "came out." At this point in the interview I often told them I was an advocate of gay rights whose research project was funded by the Gay Lesbian Alliance against Defamation (GLAAD) to study homophobia and sports. Having already established rapport and respect due to my knowledge of Rome and sports in general, all eighteen of my interviewees commended my stance. Given the history of homophobic and sexist practices in sports bars, I thought these conversations were productive and uniquely subversive of hegemonic masculinity.

It is not clear if this unprejudiced attitude is representative of the larger Rome audience. In fact, on *The Jim Rome Show* Web site (www.jimrome.com), there were sixteen pages of self-described clones that passionately opposed Rome's antihomophobic takes. Here's an example of the deep-seated homophobia of some of the show's audience members from its message board. The email response, from a man named Roger, reveals how many who subscribe to dominant masculinity feel threatened by Rome's position (retrieved on August 11, 2001):

> My thirteen-year-old was working with me at my business today and we were listening to Rome when he takes off in his "gay defender" mode. My son looks at me and says, "Dad, what's wrong with this guy? He thinks homosexuality is normal?" Well, clue-in Romey—it IS wrong: morally, and in every other way. Why you pander to this group is beyond comprehension.

These heterosexist remarks reveal that the clones are positioned along a continuum of competing ideas about contemporary masculinity. Some clones take up more pro-feminist and liberal stances while others, such as Roger, embrace more conservative and patriarchal masculine discourse, another indication of the multiplicity and instability of modern manhood. Furthermore, Roger's comments illustrate the conflicting ideas that currently exist in the United States regarding homosexuality. These conflicts are reflected in the division between people who support gay marriage and others who are passionately opposed to the legitimacy of same-sex unions. Sports radio, since it often deals with issues of sexuality, can serve as a site for productive discussion of sexual politics. Male sports fans are forced to confront and take a position on homosexuality and face their own homophobia. In addition, gay media activists and supporters (such as Jim Rome) can use the radio airwaves to discuss how gays and lesbians are excluded from two public spheres of life: marriage and the sports field.

One person was skeptical of Rome's antiheterosexist posture, viewing him as "hypocritical." Sam, a graduate student, stated that while he still listened to sports radio due to his love of sports, he was offended by Rome's "have your cake and eat it too" position:

> A contradiction! He's totally a hypocrite. Here is a so-called gay advocate on one breath and in the next breath; he refers to the LPGA as the "dyke" tours. And remember, he's the guy who got famous for calling Jim Everett "Chrissie." Plus, he panders to athletes and celebrities such as Jay Mohr. I was listening to Romey in May when Mohr called Mike Hampton [a baseball pitcher] a "gay Curious George." Rome laughed at this and lauded Mohr's brilliant humor. He's not progressive. If he was, he would confront homophobes. He's just another macho guy who's using social issues and controversy to gain market share, profits, and more radio affiliates. And his clones are just a bunch of Neanderthal, slow-witted men who have nothing better to do than listen to sports talk all day.

Sam's comments reflect his aptitude in adopting a critical or resistant position in relation to the text of *The Jim Rome Show.* During my interview, Sam made it clear that he believed his educational status provided him the capacity to be analytical, gender-sensitive, and skeptical of sports radio. One way of theorizing how and why my research subjects and other fans of sports radio (such as Sam) adopt different discursive standpoints in relation to the text of *The Jim Rome Show* is to draw upon Pierre Bourdieu's (1984) idea of cultural capital. Bourdieu coined the term "cultural capital" to describe the acquisition of social status through cultural practices that involve the exercise of taste and judgment. Originally applied in the context of educational research in France, the concept has been extended to a wider field of social distinctions in which tastes and aesthetic judgments function as markers of social class.

For example, a key distinction that Sam underscored was his access to graduate education, a form of cultural capital that served as the chief factor in giving him the ability to be critical of sports radio. The clones are described by Sam as "Neanderthals" and "slow-witted," and he complains that sports radio has little aesthetic value. Criticizing Rome's show serves as a way for Sam to reaffirm his privileged class and educational position and to distinguish himself from the sexist and lower-class clones. This position also reproduces the dominant idea that working-class people do not have access to an analytic mode of talk or their own forms of cultural capital.

7 7th Inning

WHERE EVERYBODY KNOWS
YOUR NAME

The chapter title contains part of the lyrics of the theme song from the hit television series *Cheers*. The characters in the cast were regular patrons of the bar, Cheers—a place away from home or work where you can meet old Norm, Cliff, and Frasier, see old friends and make new ones, and interact with people who share something with you other than a house or job. The sports bars I hung out in were very much *Cheers*-like environments where community was produced, and people knew your name.

This chapter highlights some of my experiences in various sports bars and the sense of group identity that occurred. This chapter also includes an in-depth interview with someone who regularly calls sports talk radio, a young man named Lou (who hung out in a sports bar on a regular basis). I met Lou at a Sacramento sports bar I visited every Monday night during the 2003 National Football League season, where the local sports radio station hosted a weekly *Monday Night Football* party. The events drew a large number of faithful fans of sports radio, which afforded me easy access to both fans and staff. Frequenting a sports bar every week for fourteen weeks offered me the opportunity to observe and participate in a particular membership community in the context of public sports viewing. I also got to eat a lot of chicken wings. In addition to describing and theorizing about the group interactions at the sports bar, I include my in-depth interview with Lou, which explores some of the reasons he had for investing so much time calling into and listening to sports talk radio programs. In particular, I examine how sports radio helped him navigate some of the complexities of contemporary masculinity and commercial culture.

As stated earlier, sports talk radio produces a speech community that fulfills the needs of men to meet in a "third space." Since many loyal fans of sports radio listen privately to *The Jim Rome Show* in their cars, the community produced is an imagined identity with little or no contact among its members. However, sports talk radio stations often sponsor events (such as a Super Bowl party) in their respective communities. These events, usually held in sports bars, often bring together local sports talk hosts and local athletes along with sports talk radio fans. Often, sports fans are able to meet sporting celebrities and other loyal sports talk radio listeners and callers. These public events, in addition to marketing the local sports radio station and its advertisers, help overcome the limits of the imagined community of sports radio and serve to legitimate fan behavior by serving participatory and social needs—third place desires—for sports talk radio aficionados.

This chapter also includes fieldwork I conducted at two "tour stops" I attended in Detroit and Sacramento. In contrast to my individual interviews with sports talk fans at sports bars, where counterhegemonic moments were apparent, at tour stops I observed performances of hegemonic masculinity almost exclusively. Many of the fans I interacted with at the tour stops appreciated the opportunity to bond with other clones, to tailgate, to be in a group context, and to consume large amounts of alcohol, which produced fairly laddish displays of masculinity.

CHEERS TO MONDAY NIGHT FOOTBALL

An ongoing community event that I regularly attended was sponsored by the local Sacramento (my hometown) sports radio station, Sports 1140 AM, KHTK. Each Monday night during the National Football League regular season, KHTK held a *Monday Night Football* party at a local sports bar/restaurant. At these events, a KHTK sports radio personality would host a get-together where local fans and sports radio listeners could gather, watch the football game, and meet KHTK celebrities. I decided to attend these events and was present every Monday night from September 10, 2003, to December 16, 2003. Attending the *Monday Night Football* party each week allowed me to meet and interact with both sports radio personnel and their listeners. Through participant observation, detailed field notes, and some informal interviewing of the participants, I wanted to explore further this imagined community of sports fans and investigate how sports radio fits into their daily lived experience. I was also hoping to meet regular callers and conduct in-depth interviews about why they call and who they are outside of the sports radio world.

The site for the KHTK Monday night parties was a Mexican restaurant located adjacent to the Sacramento River. The restaurant draws both women and men, along with children and families, as food is underscored. The restaurant has two large rooms, an upstairs, and an outside patio for dining on the river. A large television set dominates one end of the main dining area and smaller television sets are suspended in all corners of the restaurant. Local sports team paraphernalia (Sacramento Kings, Oakland Raiders, San Francisco 49ers) are abundantly displayed on various walls and over the bar area. Though sports are always playing on the television sets, management described the setting "as more a sit-down Mexican restaurant, rather than a true sports bar. People mainly come to eat; watching sports is secondary." This was definitely true during my fourteen-week participation observation; many families seemingly uninterested in the game were in other areas of the establishment, dining on Mexican cuisine. The central dining room, seating over 100 people, was reserved for the weekly KTHK party. As the dining became transformed into the sports radio party, it became increasingly gender segregated, dominated by males wearing various athletic jerseys. Drinking definitely superseded eating during the game, which confirms Gantz and Wenner's (1995) thesis that when drinking is foregrounded, the bar becomes more gender separate.

The large television became the primary focus as the game began and the patrons turned their attention to the game. The sports radio host, accompanied by two rather scantily clad women—the Sports 1140 sports babes—sat just below the television and introduced himself to the large crowd watching the game and would periodically, during breaks in the game, give away such prizes as tickets to an Oakland Raiders game, KHTK T-shirts, and free sausages. The events drew a fairly regular crowd of loyal fans. By week three, I recognized a number of patrons and knew some by their first names. It was clear that a sense of membership and community participation was established by these events. As the game developed, many men from various tables would talk about it and give their opinions of various players. When local teams such as the Raiders played, cheering and approving glances to fellow Raiders fans were commonplace. There was an unofficial sense of friendship, some bond that linked us to each other, me included. As Eastman and Land (1997) note in their research at sports bars, "membership feelings are enhanced by watching sports with other men, and the public context supplies moments of acknowledgment and acceptance by other male fans" (p. 165).

I found it easy to introduce myself to strangers because of my knowledge of sports; sports talk served as an acceptable ice-breaker. Crabb and Goldstein (1991) state that "sports are frequently considered safe topics of conversation, particularly among American males" (p. 367). Since the space in the central dining area often became crowded, I would invite men who were standing up watching the game to join me at my table. I often would use conversation about the game to break the ice and establish a comfortable context. I would then ask assorted customers why they attended the KHTK event. One "regular" said that he feels "at home in a sports bar. The people here are serious sports fans and serious sports talk radio fans like me. I feel like this is my home away from home. I try to come to every *Monday Night Football* game." The majority of people I interviewed said that they were loyal fans of KTHK and *The Jim Rome Show* (KTHK is the local syndicate for the Rome show) and enjoyed meeting the KHTK personalities and other talk radio listeners. Historically, talk radio fans have been represented as deviants in such films as *Talk Radio* (1988), *Pump Up the Volume* (1990), and *The Fisher King* (1991). John Caughey (1994) notes that popular culture fans—soap operas, science fiction, or sports—are unfairly marginalized as pathological and asocial. Social interactions with sports talk hosts and other audience members at events sponsored by sports radio stations may serve to legitimate and depathologize their sports fan status.

I enjoyed my fourteen Mondays at the sports bar and genuinely felt a sense of community among the hardcore sports fans. The KHTK-sponsored events allow the unification of strangers—sports radio fans—within a public social context. The *Monday Night Football* events provided an excuse for a gathering of the imagined community of sports radio fans, but the public event may have been of greater value than the actual football games. Just as major sporting events—such as the Super Bowl or the World Series—are cultural activities, gathering to watch the game in the public space of a sports bar is a form of cultural production and consumption. Over the weeks, I recognized many of the "regulars"—men who attended every week—and got to know many by their first names, a site "where everybody knows your name."

As the weeks passed, I became fairly well acquainted with two "regulars," Eric, a thirty-one-year-old white male, and Mauricio, a thirty-two-year-old Latino male. They are friends who work as construction employees and are faithful listeners of *The Jim Rome Show*. Eric and Mauricio both described their love for "Rome's tell-it-like-it-is approach" and his "sick humor," which includes demeaning Michael Jackson, O. J. Simpson, and Tonya Harding. Mauricio said that he once called sports radio (never Rome, though) to compete in a KHTK call-in contest (which he won for correctly answering a sports trivia question).

Wenner (1998b, p. 305) notes that sports viewing provides an intimate experience for many men and, due to its association with alcohol consumption, gives men "permission to open up." This was definitely the case in my interactions with Mauricio and Eric; by their third pitcher of beer they were much more open and expressive. As the alcohol flowed, their behavior changed, too. They often hugged each other and occasionally slapped each other's butt when their favorite team scored a touchdown. In *The Arena of Masculinity,* Brian Pronger (1990) argues that sports allow men to exclude women from their all-male environment and permit them to play with each others' bodies without suffering the stigma of being homosexual. Often after hugging each other, they would ironically say, "fag."

During their homoerotic displays, Mauricio and Eric would make it very clear that they were heterosexual. In fact, Eric marked his heterosexuality by bringing his girlfriend, Susan, to the event each week. Susan appeared very disinterested in the games and the male gender displays; she mainly kept quiet and was usually not part of the conversation. I wondered why Susan was there, unless it was to diffuse any homoerotic tension (as did the KHTK sports babes who walked around the bar awarding prized to patrons) that existed between Mauricio and Eric. The triangle between Mauricio, Eric, and Susan recalls Eve Sedgwick's (1985) ideas about homosocial desire. From her viewpoint, women are seen as exchangeable property for the cementing of bonds between the men who "possess" them. Sedgwick's central idea in *Between Men* is that heterosexual relations can be strategies of homosocial desire. That is, heterosexual relations ultimately may exist to create bonds between men; such bonds, she continues, are not detrimental to a concept of masculinity but actually definitive of it. Sedgwick identifies the strategies of homosocial desire as "erotic triangles," relationships in which there is rivalry between two active members (often, but not exclusively, two males) for the attentions or affections of a "beloved" third. Often, Mauricio and Eric would argue over who was more handsome and who knew more about sports as a way of impressing Susan.

In addition to debating who was more attractive, betting was another important social interaction between them and many of the male patrons; betting seemed to be a part of their performances of manhood. For example, Mauricio and Eric would often bet on plays or points while watching the game, with the loser having to pay for the next beer. Many men I interviewed stated that the main reason to listen to sports radio was to learn about the odds and other particularities of an upcoming game. This information would be used to place a bet with their (illegal) bookie. Sports talk radio, including *The Jim Rome Show*, often have odds-makers and fantasy football experts (fantasy football is a betting game in which winning or losing is based on the performance of individual football players) to give tips on placing bets on sporting events. Cosgrave and Klassen (2001) suggest that gambling, once viewed as sinful and immoral, has been legitimated by the state and various social and economic forces since the late 1960s. The legalization of gambling—due to economic deregulation, neoliberal philosophy, and new sources of revenue for the state—has contributed to the social acceptance of gambling activity, "and for many citizens, lottery players and sports betters for example, gambling has become a routine aspect of everyday life" (Cosgrave and Klassen, 2001, p. 3). Sports talk radio has had a major influence on legitimating gambling as a fairly harmless source of leisure and entertainment.[1]

In addition to listening to sports radio for "gambling tips," Eric thought that listening to sports radio gave him "a voice." While watching the October 6, 2003, *Monday Night Football* game between the Colts and the Buccaneers, Eric said that he could not afford to attend games any more: "They have outpriced the blue-collar guy. That's why I come here. For the cost of a few beers, I can watch the game with other fans. The tickets, the parking, the beer prices are too much at a Kings or Raiders game . . . can't afford it." Eric passionately said this while wearing a NASCAR baseball cap. He continued: "Sports radio gives me a place to share my opinions and vent some of my frustration at the way athletes are nowadays." Eric went on to utter his frustrations at how "rich teams like the Yankees" are able to "steal our players" (the Oakland As, who are considered a small-market team that cannot afford to pay high salaries for players). With more beer, Eric expressed escalating anger at "rich team owners" such as Al Davis (Oakland Raiders) who moved the team from Oakland to Los Angeles in the 1980s.

Eric was expressing a fairly common set of feelings in sports radio—fan disenchantment with: (1) the number of sports franchises that have moved from city to city for economic reasons, and (2) the super-rich teams that dominate the elite player talent pool. Fans are increasingly angry at the capitalist indiscretions in the sport world and ventilate their pain on sports radio programs. Often, this disillusionment has inspired fan rebellion such as that after the 1994 major league baseball strike, when fans were reluctant to attend games; attendance suffered in many areas. Eric's insightful comments allowed me to see how sports radio may be one of the only media-sanctioned outlets for anticapitalist sentiment in the United States. In fact, in a different context, fans' critique of the unfair economics of sports would be labeled as socialist. Yet in the context of sports radio, fans' critique of capitalism is a commonly accepted topic and may serve as a vehicle to inform people about the problems created by global capitalism. Any fan who has complained about his team's leaving town to set up in a more financially obliging city should be able to see the some of the negative effects of globalization—the deleterious effects of capital pursuing cheap labor around the world through the practice of outsourcing. Cary Watson (2002) echoes this same argument:

> Perhaps the most interesting aspect of the anticapitalist bent of so many fans and sports journalists is that it creates a fertile environment in which to educate people about the larger problems created by a capitalist economy . . . It's not a huge jump to show people [sports fans] how the capitalism that ruins their favorite team or sport can, and is, ruining lives within and outside the U.S. . . . In these red, white, and blue post-September 11 days, with media-encouraged jingoism at an all-time high, and a president and Congress that can be charitably described as corporate America's courtesans, it's somewhat comforting to realize that a rich vein of anticapitalist emotion and thought still exists in America, even if one has to go to a sports bar to hear it. Chicken wings, anyone? (p. 12)

LOU FROM LODI

One of my research questions was, "Who calls sports radio?" Based on the wave of ads you hear for penis-enlargement devices, creams preventing hair loss, and weight-reducing products, the typical sports radio caller (and listener) is a bald, fat man with a small penis. I wanted to find out who the regular callers are and to confirm or dispel this notion. Hanging out during the weekly *Monday Night Football* event afforded me the potential for meeting and interviewing some of the regular callers of sports talk radio. Each week, I would approach various men at the bar and ask them if they called sports radio. The majority, close to 95 percent, said that they only listened to sports talk. A few stated that they called once or twice, never regularly. Nobody I approached had ever been on *The Jim Rome Show*; one man, an investment banker, told me, "I tried to call once but I had to wait over an hour . . . forget about it!" as he swigged another pint of his low-carb beer and turned his attention to the game.

The process by which I engaged interviewees, in another context, would have been viewed as homoerotic and even queer. For instance, regularly approaching men, a bit nervously at times, and asking them if I could interview them, get their phone number, and buy them a beer mirrored pick-up actions at a gay bar. Only the setting and context were different; there were the hyperfeminine, heteronormative, KHTK "sports babes" in the background, along with my knowledge of sports radio to reassure the strangers about the heterosexual nature of our encounter. Plus, the men were watching sports and wearing Oakland Raiders, Sacramento Kings, or San Francisco 49ers gear, not dancing to house music and wearing trendy outfits. Nevertheless, it was obvious to me that the interactions in the sports bar and relations at a gay club were not as distinct as many homophobic sports fans might like to imagine.

Eventually my consistent approaches paid off as I met the famous "Lou from Lodi" (name and place changed to preserve confidentiality), a regular caller of local sports radio. Many of the regular callers of both national and local sports radio take on an iconic status within the world of sports radio and are often known by their nicknames such as "Positive Dave," "Republican John," "Gentleman Ken," or "The Sign-man." In fact, *The Jim Rome Show* recognizes the contributions of creative and "regular callers who have managed to rise above mere clone status." Rome salutes their contributions to the show by marking them as "legends" whose "contributions or jerseys hang from the rafters." The legendary clones included "Silk," "Jeff in Richmond," "Slam Man," and "JT the Brick" (who now hosts his own national show).

"Lou from Lodi" has reached this legendary status in Sacramento. I recognized his voice right away, but expected him to be a bit older than his twenty-one years. Lou was wearing a 49ers jacket, standing up near the large screen television watching the November 3 Denver Broncos–New England Patriots game. While talking about the game with Lou, we heard the KTHK sports host announcing to the rowdy crowd that "Lou from Lodi" was "in the house." The congregation of sports radio fans wildly cheered; it was apparent that most present knew Lou from his daily "takes" on KHTK. Lou seemed to revel in his celebrity status as he waved to the crowd.

"Boy, you're quite famous, Lou," I noted. I then went on to describe my research project and my interest in interviewing him as a regular caller of sports talk radio. My desire to include him in my study only added to his star status. He agreed to be interviewed as long as I bought him a Mountain Dew.

As we walked over to the bar to get a bit of privacy, I asked, "Why do you listen to sports talk radio?"

He shrugs, sipping on his soda. "Because I'm in the car a lot driving to school."

"How many hours do you listen a day?"

"About six hours a day. I listen in the morning and then I can't miss Rome."

"So you listen other than when you're in the car?"

"Oh yeah," he smiles mischievously. "I sneak it at work, too!"

"What do you like about Romey?" I inquire.

"I love his takes and his sarcasm."

I was impressed with Lou's deep resounding broadcaster's voice. In fact, Lou is earning a degree in communications at the local university and is interested in working in the sports talk radio industry. Lou, who identifies as "Portuguese and Mexican," also works part-time as an office assistant and does some disc jockeying at a local hip-hop club.

"So, back to business, Lou, do you listen to music or other radio programs besides sports talk?"

"Nope, I hate that Dr. Laura. I like Howard Stern and Rush Limbaugh, but they're on the same time as Romey."

"Why sports talk over Stern or Rush?"

"I don't know. Sports talk is just fresher. I'm a big sports fan and the callers are more interesting."

"No NPR?"

"No way. That channel is for liberals!"

Lou goes on to say that he's a dedicated Republican and a big Bush supporter. While our political affiliations are very different, much of how we spend our days—consuming sports media—is similar. I, like Lou, listen to sports radio while in the car, rarely NPR. When home I, like Lou, tune in to ESPN's *SportsCenter* or Fox Sports Net's *The Best Damn Sports Show Period* (when not watching live sporting events). We have a bond through sports, despite my opposition to the Iraq War and my intense desire to get Bush out of office.

When Lou states his political affiliation, it is halftime and the KTHK host is introducing a soldier who has recently returned from Iraq. The crowd and Lou cheer as the host gives him a KHTK T-shirt and thanks him for "protecting us all back in America."

I ask Lou about the Iraq War while thinking to myself about the link between sports and nationalism.

"Bush had to go and finish the job of his dad."

Lou's comment on George W. Bush's needing to complete his father's "job" was fertile territory to explore ideas about nationalism, sports, psychoanalysis, and patriarchy. I wondered to myself while sipping a margarita: Was Lou implying that Clinton feminized the United States (similar to traditional ideas that mothers feminize sons if they do not separate at an early age and connect to father)? Was he suggesting, informed by a warrior masculinity narrative, that George W. needed to reinitiate the country back into manhood by following through with Bush Sr.'s militaristic wishes? My therapist self interprets Lou's remark through a psychoanalytic lens. Perhaps it is the setting; possibly I am overanalyzing and the margarita is going to my head; maybe I am being too "therapeutic" here. I decide that a discussion on Iraq will likely lead to a counterproductive debate that I am too tired to get into. And so I switch the subject back to a subject with less con-flict—calling sports talk radio shows.

"You remember your first call?"

Lou scratches his shaved head, and remembers. "It was just a cool thing to do. I called during a quiz show when I was fourteen years old. I didn't win, but I got a rush out of it. I've been calling ever since."

"What's been your experience calling Jim Rome?"

"I have only done it three times. I thought it would be cool to fea-sibly be in the running for the Huge Call of the Day. I felt I could do better than most of the takes."

"What was your take? Did you write it down?"

"Yeah, I wrote it down because I didn't want to get run."

"Did you?'

"No," Lou proudly exclaimed. "In fact, he racked me! But I didn't win the Huge Call."

"What was your take?"

"Oh, it was something about Terrell Owens. I thought clones should get off his case about his celebrations in the end zone."

"You're a big Owens fan?"

"Oh yeah!" Lou then takes off his jacket to reveal a 49ers jersey with Owens's number and name.

"Why do you think he gets so much shit from callers and fans?"

"I don't know. He's just his own man. He's real."

I decide to approach the issue of race.

"Do you ever think race is part of the reason why white fans don't like Owens . . . like he's too threatening?" While discussions of race are typically fraught with tension, I felt comfortable in broaching this subject. Since race is often discussed on Rome's show and other sports talk programs, it was not an unfamiliar subject to Lou; sports talk radio served as a vehicle to talk about the politics of race.

"Maybe . . . well, I don't know," Lou responds tautly. "I think that sports is pretty color-blind."

I go on to tell him my thoughts about race and sports while Lou, who is very hyper, listens with a certain calmness and interest.

"I haven't thought about it that way, but you're right. The players that get the shit about celebrating are all black. Maybe you're right on, dude. Nice take!"

We both laugh as I order him another Mountain Dew. Through the speech style of sports talk, perhaps we had a moment of critical dialogue about race. Maybe?

"So, back to the Rome show, Lou, do you mind waiting so long on hold?"

"Not really. It's long . . . an hour or so . . . but you're listening to the show so it's not so much a waste of time. But I don't call him much anymore. He doesn't take too many calls now. I like his interviews anyway. It's a lot better when he doesn't have too many calls. There are just a few callers who are really good, like Silk."

"Do you consider yourself a clone?"

He is offended. "No, clones are like sheep!"

"Like the ditto-heads on Rush," I ask referring to the loyal listeners of Limbaugh's hate speech show.

"Exactly. They don't have a mind of their own."

I wonder to myself if Lou would also eschew the identity of "clone"

if he knew the meaning of the word in 1970s New York or San Francisco. In the 1970s, and before the AIDS crisis, there was a radical
shift in gay male culture, as a male homosexuality materialized that
embraced a more traditional masculine ethos. The gay male who performed this traditional masculinity was known as a "clone," a muscle-
bound, sexually free, hard-living Marlboro man, who appeared in the
gay enclaves of major cities, changing forever the face of gay male culture. Levine and Kimmel (1997), in their ethnography of the gay
clone, describe clone culture as "gender conformist," reproducing
many of the features of hegemonic masculinity. Although the gay
clone had different priorities than the sports clone ("disco, drugs,
dish, and dick," according to Levine and Kimmel, 1997, p. 1), their
performances of manhood—features of manhood that Kimmel and
Levine critique, whether practiced by gay or straight men—may have
more in common than sports talk clones would like to acknowledge.
However, I decide not to disclose the history of the homosexual clone
with Lou. Instead, I ask about his calls on local talk radio.

"So, you mainly call KHTK?"

"Yeah, I like the back-and-forth talk on KTHK. You get more time
to take and it's more interactive. You can share your opinion with the
local guys and they don't cut you off."

I agree with Lou. In my analysis of both local programs and national shows such as the Rome show, there is much more opportunity
for dialogue in the local programs. I have called sports talk radio twice
in my life and both times I called a local show. Rather than engaging
in a monologue with Rome and letting him decide to run or rack me,
there is more democratic interchange on local shows. I appreciate
local hosts' idiosyncratic voices and seemingly genuine interest in
hearing the caller's opinions. Despite the show's being disseminated
by huge media conglomerates, local sports radio seems to thrive on
the quirky and regional. This is not to deny the productive textual moments on *The Jim Rome Show*, when Rome uses his power to police
the boundaries of acceptability and end any conversation that introduces racism or homophobia. Yet it's been my experience that local
sports radio, while occasionally engaging in more overt forms of
homophobia and sexism, also monitors the boundaries of speech and
will not tolerate political hate talk. Haag (1996) may be correct when
she suggests that local sports talk serves as a model for what democratic activists and theorists have imagined as a template for political
and public discourse of civic society. Haag writes that sports radio is

"one in which people can speak both passionately and respectfully. They can care deeply about issues, but no one will die for expressing opinions on them—unlike those involved, in say, the abortion debate" (p. 461).

Lou is one of the models of such civic discourse. Although he refers to women as "a different animal," invoking gender essentialism, uncritically supports the Iraq War, and is fairly committed to traditional notions of manhood, he speaks both civilly and respectfully when he calls sports radio (and he is very kind and decent in person). I know. I have heard him several times on the radio and often agree with his opinions. I decide to tell him this.

"Your takes are right on most of the time."

"Thanks, man. I appreciate it. In fact, many friends and other callers will say that they liked my take on such and such . . . it's a compliment. That's why I call. I like to make a difference and mix it up with other sports fans."

"Do you have a strategy when you call?"

"Yeah, I try to make my calls not boring. I try to make them fast and exciting."

"I see," thinking to myself that in my other role as a therapist, I might diagnose Lou with Attention Deficit Disorder (ADHD).

Right then Lou starts tapping his fingers on the bar. His ADHD-like squirminess is kicking in. It's my signal to wrap up this interview.

"So, you think sports radio is a fad?"

"No, just look around this bar. Look at all these fans! It's here to stay."

Lou walks over to the KHTK host, who is mingling with the patrons as the second half begins. I think to myself that Lou is part of an important and somewhat inclusive community of sports fans, people who need connection, through sports discourse, from an increasingly isolating and economically marginalizing culture. Sure, from within certain circles in the academy, sports fandom is a culturally trivial activity. But do *not* tell that to Lou and other people I have interviewed in this project.

A COMMUNITY OF CALLERS

Sports talk radio increases its ratings during playoff times.[2] In my hometown, the beloved Sacramento Kings were in the playoffs for the sixth straight season (2004). Expectations were high as many Kings fans, including women, called the local sports station, KHTK, to express their "take" on the King's chances for a National Basketball Association (NBA) championship. A popular KTHK program is their morning *The Rise Guys* show where three irreverent hosts—"Whitey," "Phantom," and "Kevin the Rat"—blend sports talk with wacky humor and banter with regular callers such as Republican John, Isaac, Thomas, and Gentleman Ken. The banter between the callers and *The Rise Guys* became more impassioned as the playoffs loomed. Thomas, who did not believe that the Kings could win a championship, suggested that *The Rise Guys* and the regular callers actually meet in person at a local sports bar and watch the playoffs. Republican John, Isaac, and Gentleman Ken all thought this was a good idea and invited *The Rise Guys* and other KHTK staff to join them. On the show after the game, Thomas and the other regular callers, along with "Whitey," debriefed the prior evening. They all talked about how much they enjoyed meeting each other in person. A real friendship was developing among the group.

Thomas's organizing the group fit with my thesis that sports radio creates community. I emailed "Whitey" after hearing about their evening, stating that I was doing research on sports talk radio. In my email, I asked if I could "hang out" with the group as their congregations illustrated my notion about the civic potential of sports talk radio. On Friday, March 14, 2004, I received a call from my friend, Debora, who excitedly said, "Whitey just read your email on the show!" I decided to call the program at that point to thank him for inviting me to their next gathering and to use the call as an opportunity to talk about my research.

I dialed KHTK's number while driving to work. The screener, "Little Joe," answered and somewhat brusquely asked, "What do you want?" I told him who I was and, before I knew it, I was on the air.

"Next caller, David," said "Whitey."

"Hi, Whitey, this is David Nylund, the guy doing the research on sports talk radio." I felt surprisingly nervous.

"Oh, David Nylund, the dude hanging out in sports bars. When we see you tonight, are you going to be wearing a white coat with note paper?"

"No," I laugh. "I will be wearing a Kings jersey. I am a regular listener of your show and Kings fan who's finishing up my Ph.D. at UC Davis. I will be carrying a pen in one hand, but most likely a pitcher of beer in the other."

"Cool," says Whitey as "Kevin the Rat" and "Phantom" chuckle in the background. "No, in all seriousness, David, I have been in radio for over twenty years and I have never experienced this before. I think it's great that Thomas and the gang are actually meeting in person. You're welcome to join us."

I am still nervous and can barely hear "Whitey" on my cell phone. My anxiety reduces me into "academic babble" about my research. I go on to tell Whitey (and thousands of listeners) about "third places," work changes, and masculinity. After my long monologue, I hear nothing on the other end. I realize that I have been run. As I turn the radio back on I can hear the end of my call and "Whitey" running me (there is a five-second delay on the radio). After the call, "Phantom" says, "Man, that guy needs a personality. Now I know why I hated college."

I immediately feel humiliated; this reminds me of times growing up when I was teased by a group of other boys because I did not measure up in some way. I call my friend Debora for some emotional support. Sports radio can be cruel. It does, at times, reproduce hierarchical relations between callers and hosts. It is one thing to intellectually theorize about it, but another completely more intense and personal experience when one is actually the caller at the other end of the power relation.

After the support I received from Debora, I decided to follow through and meet up with Thomas and the gang later that evening. When I arrive, I see a reserved table for KHTK. Soon, an African American male wearing a Kings jersey sits down at the table. I go over and introduce myself, hoping he is one of the callers. The man introduces himself as Isaac. Right then, a middle-aged white male, accompanied by a woman, joins us. Isaac introduces me to the man, who is Republican John. Republican John is a very interesting caller. In addition to his interesting "takes" on the Kings, he often espouses his conservative political opinions on the air. I decide to ask him about his politics.

"I'm not the kind of conservative that is it part of that Moral Majority shit. I don't care what two men do in the privacy of their homes," Republican John shouts out as the bar is getting louder with excitement as the Kings/Timberwolves game is about to start.

"You're more of a libertarian?"

"Yes, exactly." Republican John tells me a bit more about his life. He is forty-four, owns a swimming pool business, and lives with a girlfriend who has children from a previous relationship. While working, he flips back and forth from sports talk radio and political talk radio.

He calls sports radio almost daily. He states that he is a social libertarian and fiscal conservative. Throughout our conversation, he checks in with his partner and is holding her hand. I ask her (known as Mrs. Republican John on the radio) about Republican John at home.

"He is very kind. He shares in the housework. And he's very good with my kids . . . like they are his own."

"So, some of the macho stuff on the air is not who he really is?" I ask.

"No, not at all. That's just for show."

Republican John, who is attentively listening, adds his perspective on gender. "I think real men should treat women with respect."

"Do you think sports talk radio, however, glamorizes disrespect of women? For instance, I took a woman friend to a tour stop and she was harassed there. It seems that sports radio may not be a safe space for women?" I rhetorically ask.

"Well, that's bullshit. No man should touch another person's property," Republican John espouses, invoking a benevolent patriarchy that reminds me of some of the discourses that inform the Promise Keepers.

I give Republican John a release of information form at that moment, which he carefully reads and signs. He then asks about my thesis on sports talk and I tell him my thoughts on how sports talk radio can help create community and temporarily break down barriers of race and class between men. Republican John agrees: "You are right on. I would not have anything in common with these fools [pointing at Isaac, along with Thomas and Gentleman Ken who have joined the table. All three of the men that Republican John is referring to are African American] if it wasn't for sports talk radio. They are my friends."

Thomas enters our conversation and agrees. "In fact, we have given Republican John honorary brother status!" The entire table and I laugh, including "Whitey" who has now joined the group.

It's getting close to game time so I quickly interview Gentleman Ken—a regular caller of KHTK and *The Jim Rome Show*. Gentleman Ken, a forty-five-year-old male, married with two children, works with severely emotionally disturbed children and is earning his teaching credential. His wife is not present. "She knows that this is my stress relief." He looks at the group and says, "I think what we have here is very cool, a real bond. Sports radio is great. On the air you can act crazy, but it's better to meet and interact in person." Gentlemen Ken has been the caller of the day six times on *The Jim Rome Show*, including winning the contest recently due to his "take" on the King's forward Chris Webber. "I like Romey, but local sports radio is better. There is more interaction between caller and host. You can have a real conversation."

Thomas, who is listening, adds: "This is great, what we have going here. Sports can bring people together."

"Why do you call sports talk?" I ask.

"I like to mix it up with *The Rise Guys* and create controversy. I just love sports." Thomas's attention turns to the Kings game. I decided to thank the group and buy them a pitcher of beer. As the game proceeds, the alcohol is flowing. Their bonding takes the shape of stylized hyper-masculine talk—posturing, jokes, and put-downs. I, as part of the group, engage in the same discourse. An outsider may view this style of talking as crude and intimacy-disabling. Yet it seems that much of the talk is a self-conscious parody of masculinity. There is real connection between these men underneath the posturing and masculinist speech. Debora, who had joined the table at halftime, also feels included and accepted. The space created here is a great good place, "a third place."

The next time the group met was the following Wednesday, which was game seven of a tightly contested semifinal series between the Kings and Minnesota Timberwolves. Unfortunately for Kings fans, the Timberwolves won 83–80 with Chris Webber's desperation three-point attempt rimming out. The next morning on *The Rise Guys*, several inconsolable Kings fans called the program to ventilate their frustration at another disappointing season. *The Rise Guys* acted as counselors to the brokenhearted callers, offering grief counseling, anger management, and even sex therapy. Yes, one woman called the show to express her glee that that season was over. "Maybe my boyfriend will have sex with me. We haven't done it since the playoffs!"

In addition to offering a communal function, sports talk radio hosts can serve as psychotherapists for fans who become emotionally invested in their team. When one's team loses, particularly in a playoff, passionate fans experience depression and aggravation. Sports radio can be a place to ventilate these feelings in a community context. I remember calling sports radio in 1996 after another disappointing Detroit Red Wings season. Calling the show was weirdly healing for me; just having a fellow sport fans hear and relate to my pain was therapeutic.

Thomas, Gentleman Ken, Isaac, and Republican John all called the program the day after the King's defeat. In addition to expressing their sorrow, they were all sad that the end of the basketball season meant no more meetings with the sports radio community they had created. Thomas already had plans to get together for the football season. I plan to continue meeting with them at various sports bars—less as a researcher and more as a friend and fellow fan.

AMONG THE ~~CLONES~~ HOOLIGANS

Tour stops are rowdy public events at which Jim Rome rewards a particular city and the local sports talk station (a Rome syndicate) with a live two-hour show, interviews local athletes, and entertains the clones with some of his comedic monologues. In contrast to just listening to the show on the radio, clones are given the opportunity to meet and socialize with other clones (as at the sports bar events), providing these fans with "a greater sense of connectedness and membership in a group than watching in private" (Eastman and Land, 1997, p. 165). Such events allow the imagined airwave speech community to come to life and present the opportunity to bond with other clones. The bonding usually begins several hours before the show as fans arrive early for tailgating.

I attended the Sacramento tour stop held at Arco Arena on September 21, 2002, with a good friend, Debora, and two male cohorts. Attempting to "blend in" with the clones, we brought our own tailgating essentials: beer, chips, and brats. As we walked through the parking lot, the attendance of men compared to that of women was overwhelming; the few females appeared to be with their boyfriends or husbands. High levels of alcohol consumption along with generally raucous behavior were the central activities in the parking lot. The public drinking reminded me of adolescence and one of its central rites of passage: drinking large quantities of alcohol in a public space with one's peers. Witnessing the rowdiness and drunkenness at the tour stop brought back memories of my adolescence and early adulthood, where proof of holding one's liquor was proof of one's manhood and drinking less risked peer rejection and feminization. Burns (1980) writes about this adolescent male ritual:

> He and his fellows must participate in whatever masculine activities are open to them to assert themselves as men. This is accomplished by "being rowdy." When young males are together, their conversations are filled with references to street fighting, acting "tough," and "being rowdy"—activities that usually involved large amounts of beer. (p. 280)

Even though the clones in attendance were adults, not teenagers, the performances of masculinity were a reproduction of what Burns describes. Tour stops and other homosocial drinking rituals may be a way for men nostalgically to reclaim their adolescence and youth—a means of escaping the responsibilities and pressures of adulthood and domesticity.

It was quickly obvious that it was unlikely I could interview clones in the context of tailgating and rowdy crowd behavior about some of the fissures and ambivalences of masculinity. The group context and the unruly male performances made it clear that tour stops were, as Maureen Smith noted in her fieldwork at an Oakland tour stop, a primary site of the "reproduction of hegemonic masculinity through the act of male bonding in a sport setting; further, the inclusion of alcohol introduces comparisons between sports bars and tour stops, but also tailgating and tour stops" (2002, p. 6). I decided to switch my method to participant observation and detailed field notes.

While observing the happenings, Kevin, a clone in his early thirties wearing a Raiders jersey, approached us and offered us beers. Before I could ask him any questions about sports talk radio, the inebriated Kevin began talking about how "awesome" Rome was: "I love Romey. He is God." Kevin had been listening to *The Jim Rome Show* since the infamous Jim Everett episode seven years earlier. "I love his fresh takes and his not-give-a-shit attitude . . . he's a real man," Kevin said, venerating Rome's aggressive masculinity. As we walked from the parking lot to the arena, Kevin joined us, bragging about his previous career as a rock musician and his appeal to women. As the show began, Kevin sat in the row behind us.

Most of Kevin's attention turned to my friend, Debora. However, I was distracted from Kevin's flirtatious behavior due to the rowdy crowd of 12,000 clones standing and cheering as the lights were lowered. Jim Rome, wearing all black, appeared on stage, pumping his fists while the Guns and Roses tune "Welcome to the Jungle" blared through Arco Arena. I felt a dose of adrenaline run through my body as Rome danced around the stage as if he were in a mosh-pit. After several minutes of music and cheering, Rome addressed the crowd: "What's up, Sac-Town!" The crowd was in sync with Rome, cheering his every movement and word as if he were a rock star. Several women then began exposing their breasts, as if it were Mardi Gras, which threw Rome off for a moment. "Nice," Rome said in reply to the breast uncoverings, which led to him talking about how he hopes his daughter grows up to expose her "knockers at a tour stop."

After the crowd settled, Rome presented his typical diatribe of bashing Tonya Harding, Michael Jackson, and O. J. Simpson., referring to strippers as "skanks," and talking about getting blowjobs (referred to as "hummers" by Rome) from prostitutes while golfing. Rome's progressive stances on race or sexuality were invisible; this was not the site for such discourse. Rather, this was a place for Rome and his clones to engage in laddish displays of manhood that resembled more a World Wrestling Entertainment match (WWE), which tends to be a celebration of bravado and violence. In fact, several fights began breaking out throughout the arena, mainly between 49ers and Raiders fans. There has been a long-standing idea that Western European (mainly British) soccer hooligan crowd violence is much greater than North American sports crowd violence, and that U.S. and Canadian sports spectators are almost uniformly peaceful. However, Eric Dunning has noted that this idea is "a myth that North American sports behavior, far from contradicting American 'exceptionalism' in this regard, is fully consistent with the violence of the USA" (2002, p. 159). Witnessing the violence in Arco Arena confirmed Dunning's position. Even though Rome attempted to stop the violence by pleading to the crowd, *The Jim Rome Show* was accountable for helping create the context of violence through his hypermasculine speech and encouragement that clones arrive early and drink.

While listening to Rome and observing the crowd, I was increasingly caught up by the excitement and fanatical power of the crowd, which distracted me from what was happening right next to me. I was particularly absorbed when Rome interviewed the Sacramento King's owners Gavin and Joe Maloof. Rome referred to the Maloof brothers as "the ultimate men . . . they own a beer distributorship, a professional sports team and a casino [the Palms in Las Vegas]. All they need to really be the definitive men is to own a strip club!" As I became immersed in Rome's hypermasculine speech, I was oblivious to Kevin as he began verbally harassing Debora by making sexually suggestive and even aggressive comments. Even worse, I was not aware of Kevin's following Debora into the bathroom in true stalking-like fashion, telling her that he "would not bring a woman to such an event." Debora was understandably relieved when the event was over, sharing afterwards how the entire event reminded her of a big fraternity party that is unsafe for women. The experience left me feeling guilty, even questioning my own manhood and ability to protect my friend. It reminded me that despite some fissures in its presentation of traditional manhood, sports radio is still firmly bound up with the production of hegemonic masculinity—there are limits in the productive potential of this genre.³ It also reminded me how sports talk, while creating intimate moments, can also direct men to be unaware of the costs of such fandom to people who are not typically hailed by such discourse—namely, queers and women.

The Detroit tour stop, July 26, 2003, was a carbon copy of the Sacramento event, both in the show's content and the pre-show tailgating, with the exception of dress: the Raiders, 49ers and Kings gear was replaced with Detroit Tigers, Lions, Redwing, and Pistons gear. Otherwise, the brats, beer, trucks, and SUVs were the same. I attended the Detroit occasion with my younger brother's friend, who is a long-time fan of Rome. We arrived at 10:00 A.M., four hours before the show, and walked the parking lot talking to clones, including the famous caller, "Silk," who was a celebrity in the parking lot, as many wanted his autograph or to be photographed with him. Each person we met was drunk—probably over fifty men (we talked to two heterosexual couples)—and was proud to claim the identity of "clone."⁴

While the rock group *Grinder* (a band that includes league singer Darren McCarty of the Detroit Redwings) played heavy-metal, we asked the clones about their listening habits and why they listened and what they thought of Rome. Most listened to Rome while driving in their car, with several salesman stating that the show got them through many grueling road trips. Their replies in regards to why they listened to Rome were predictable: "Rome is God"; "He's real, tells it like it is"; "He's the man." Many clones confessed that they loved the King of Smack's "nonsports" topics. With the exception of one man who liked Rome's criticism of "wife beaters," most commented on Rome's crude humor on nonsports topics, not on his political commentary. "I love that he talks shit about people who deserve it, like strippers or Tonya Harding," said a drunken shirtless clone. Several rowdy clones loved Rome's recent "takes" on men who play a game in Las Vegas called "Hunting for Bambi"—a game in which men play huge sums of money to "hunt" naked women in the desert and shoot them with paintballs. "Man, I wish I had the money to hunt for Bambi," said one Rome fan, while his friend exclaimed, "Nah, man, if I had that much money to burn, I would buy a bunch of fucking hookers." Everybody laughed, except me. Even though I was uncomfortable with the speech, I did not want to explicitly challenge their sexism due to the group context and their public drunkenness. Yet my silence was noticed by some in the group and created a moment of awkwardness until one man cracked another joke about sex workers. The talking became increasingly misogynistic and homophobic as the show start-time approached, and the beer consumption increased as some fighting broke out; it was very clear that this space served as a refuge from women and marginalized men. In fact, many clones said the tour stop allowed them to get "away from the wife."

Maureen Smith's (2002) account of the Oakland tour stop resembles much of my experience of the Sacramento and Detroit events.[5] Smith also had to contend with this masculine site as a woman researcher. I quote her at length here as she describes one of her experiences in the Oakland Arena:

> Other incidents are more disturbing. A group of Oakland Raider season ticket holders are sitting in the first two rows of the audience. They are posing for pictures with each other and when I ask them if I can take a picture of their group, they oblige, flipping me off in a gesture of solidarity. I do not interpret the middle finger gesture as a sign of disrespect, but more a sign of their group mentality (they had been drinking in the parking lot since five in the morning) . . . They [the group of Raider fans] have a Hanes T-shirt they have written on with a permanent black ink marker. On the shirt is a handwritten comment: "Let's rape KC—no lube." The Oakland Raiders are scheduled to play their divisional rival, the Kansas City Chiefs the next day. The use of the word rape and the added emphasis of "no lube" paints a violent picture of the ways sexual language has been used in sport to send messages. It also indicates a powerful example of how some men enact the lessons they learn during their consumption of sport; lessons of dominance and power. (p. 17)

Behaviors that in a work setting or other social settings would be unacceptable found reward and praise when exhibited at the tour stop. Curry (1991) reminds us that male refuge places, such as locker rooms and sports bars, are "likely to have a cumulative negative effect on young men because it reinforces the notions of masculine privilege and hegemony, making that world view seem normal and typical" (p. 133).

As I left the Detroit tour stop, I wanted to make sense of my contradictory experiences in my audience research. Why was I able to have more in-depth, even antisexist, conversations in my sports bar research? What was different about the context of the tour stop? I realized that, in sports bars, I typically interviewed my subjects individually in a private corner of the bar, whereas my interactions at the tour stop were in a group context. As noted by Kimmel (2004), men in groups have difficulty exhibiting alternative forms of masculinity; the pressure to be "one of the guys" is immense in normative, heterosexual male groups. In my personal experience, my work as a therapist, and in my research encounters, men will "open up" and express more of their preferred selves in private one-on-one interactions. Maybe the collective, laddish masculinity expressed at the tour stops was more ironic? I know my brother's friend who attended the Detroit tour stop, while enjoying Rome's vulgar humor and laughing at much of the clones' sexist comments in the parking lot, is a kind, decent man and a committed father and husband. In fact, he told with me that his wife was his "best friend," sounding like a "new man" rather than a babe-watching, self-centered "new lad." The more laddish displays of masculinity he exhibited at the tour stop seemed to be more stylized, self-consciously exaggerated, and playful; hanging out with men talking about prostitutes, getting wasted, acting lewd, rowdy, and making homophobic gestures was not a prominent feature of his actual daily lived experience. Perhaps for some the tour stop afforded men a temporary reprieve, a retreat into traditional masculinity. For others, the violence, sexism, and homophobic othering appeared central to their identity, and the tour stop helped support and reinforce their investment into hegemonic manhood and maintenance of the patriarchal social order. My audience research at the various sites confirms the notion that masculinity is fluid, contested, and expressed differently in group and individual contexts.

My analysis also indicates that the more liberal discussion of sexuality and race occurs on the airwaves rather than in person. Why? Conceivably there is something about the aural medium of radio that allows for subversive moments. Allison McCracken (2002), in her analysis of radio thriller dramas of the late 1940s, suggests that because radio is invisible, the voice cannot be visually fixed to a gendered body as in cinema, and therefore the disembodied women's voices is potentially disruptive. Likewise, the detached male voice, whether it is Rome or his clones, might create the possibility of experimenting and playing with alternative expressions of masculinity on the radio airwaves. Surely, as noted throughout this study, talk radio is a key site of hegemonic masculinity. Yet the radio allows for the detachment of voice from the visual, allowing some men to step outside the narrow confines of traditional masculinity since other men are not physically present to visually disapprove of their speech. In a group context, heterosexual men not only verbally insult each other, but often use nonverbal cues to signify disapproval if one challenges conformist masculinity by speaking from a feminist or antihomophobic perspective. Hence, the radio might be a space in which men step into alternative modes of being and speaking without the visual policing of other straight men.

8 8th Inning

A SPORTS RADIO INTRUDER

This chapter highlights a listener to and occasional caller into sports radio, my friend Joan. Joan, a passionate sports fan and athlete on the Sacramento Sirens women's football team who is openly "out" as a lesbian, listens regularly to *The Jim Rome Show* and other sports radio programs. Since sports talk radio is marketed toward heterosexual men, I was interested in how Joan—someone whose social location is not considered part of sports radio's niche and one often marginalized by the textual themes embedded in the genre—uses and understands sports talk radio. I will discuss some of the innovative strategies she uses (including irony, sarcasm, and actually calling into and participating in the show) to resist the sexist themes of sports talk radio. Joan's comments illustrate how one can enjoy sports talk radio while not internalizing the patriarchal values promoted by the genre.

Super Bowl week is always big on sports radio as callers and hosts intricately discuss every aspect of the upcoming game. On the Wednesday before the 2003 Super Bowl I was listening to the local evening show, in which the host and his callers were focused not on the game but on a halftime pay-per-view event called the Lingerie Bowl. Marketed by Horizon Productions and Dodge as an alternative to the Super Bowl halftime show, the Lingerie Bowl featured models dressed in lingerie playing a seven-on-seven tackle football game. Many callers enthusiastically said that they were paying the twenty dollars to watch, as Horizon Productions described on their Web site (www.lingerie bowl.com), "beautiful athletic women against each other in a fiercely competitive seven-on-seven game of full contact tackle football."

The celebratory talk on the radio came to a halt when the host said, "Next up, we have Joan who's in a truck. Joan, what do you think of this Lingerie Bowl?" When Joan began speaking, I immediately recognized her as my next-door neighbor. Joan went on to say that she was offended by the sexism of the Lingerie Bowl, saying, "If you really want to watch real women play football, come see the Sacramento Sirens play." The Sirens, Joan informed the host and the listeners, were the 2003 Champions of the Independent Women's Football League. Joan, an offensive guard for the Sirens, encouraged the host and his listeners to buy tickets for an athletic and entertaining "brand of football." The host was supportive of Joan's plug and actually apologized for some of "the male chauvinistic talk," agreeing that the Lingerie Bowl was "objectifying of women," pledging not to pay money to see it. After Joan's eloquent critique of the Lingerie Bowl and her plugging of "real women's football," the conversation shifted as callers went back to discussing the Super Bowl match-up between the Patriots and the Panthers. To my surprise, a few callers agreed with Joan's talk, stating that the Lingerie Bowl was offensive. Some even agreed to attend a Sirens game—a momentary fissure in sports talk discourse.

Bolin and Granskog (2003) suggest that women athletes are "athletic intruders" since they challenge the male dominance of sport. Similarly, Joan was a "sports talk intruder" that Wednesday evening by shifting the discourse and disrupting the male preserve of sports radio. In addition, Joan is an athletic intruder due to her playing a sport that is typically associated with masculine hegemony—football. Football, probably more than any other U.S. sport, reinforces male hegemony by linking such qualities as aggression, force, and violence to biological maleness, which implies female inferiority. If women play football, it must be contained, such as in the Lingerie Bowl, where the players must perform a hyperfemininity by wearing the adornments of heterosexuality—lingerie—to reassure viewers of male physical supremacy. Female intrusion into authentic, full-pad, tackle football deconstructs particular practices that are considered quintessentially male domains challenging the traditional masculine/feminine dichotomy that has been associated with sport. I have attended many Siren games and met many of Joan's teammates in social situations, and, like some other men who have also met them, have been surprised at their level of aggression, drinking habits, amount of profanity, and their overt appreciation of queerness.[1] For women's football to be legitimately discussed in the male world of sports talk radio was a sparkling moment. I couldn't wait to call Joan and congratulate her.

When I called Joan that evening to applaud her "take" on sports radio, I found out that she is a regular listener to sports radio, both local and national programs, including Jim Rome. She agreed to be interviewed about her experience listening to sports radio and consented to my using her real name. Joan, thirty-four, lives with her partner, Gretchen, and their two daughters. In addition to her full-time work in the technology field and her football playing, Joan volunteers at the Sacramento Lambda Center—a community program that serves the lesbian, gay, bisexual, transgender, queer, and intersex community through support, education, advocacy, and referral services—where she facilitates a queer youth support group. Joan has reaped the benefits of Title IX, having played sports all her life.[2] An avid Minnesota Vikings fan, Joan always wanted to play football. When the Sirens were formed, Joan was one of the first to try out. She has been a dedicated and key player for Sacramento for four years.

Joan listens to sports radio while on her forty-five minute commute to work. "It's better than Stern or Limbaugh, and plus I get to keep on the world of sports," Joan confessed during our interview in her living room. "I like Rome because his interview with athletes is excellent, he's able to get players to open up, plus his sports knowledge is pretty good. I like his in-your-face attitude and some of the takes are entertaining . . . but I have to turn him off at times."

"Why?" I inquired.

"I can't stand his takes on women's sports. He recently infuriated me with his take that no one cares about the WNBA finals. I was furious. I even thought about calling the show to protest."

"Why didn't you?"

"Well, you know . . . Rome would just 'run' me. He's the king of smack and you don't agree with him, he cuts you off. No wonder they're called clones!"

"Are there things you don't like about the show?"

"He's just pretty misogynistic . . . Hey, I know I, a lesbian, am not part of his niche. I feel like I am eavesdropping on a male-only conversation. It's kind of cool. I don't take his program so seriously. I take it more ironically or mockingly, just getting the sports information I need. Not like his clones that live on every word Romey says, thinking he's God or something. Some people who take it literally and do not see the irony and sarcasm remind me of some stupid ass who watches *Jack Ass* (an MTV program in which white, straight men perform dangerous stunts) and tries the stunt at home and gets injured."

Joan's strategy of listening to Rome fits with Michel de Certeau's (2002) notion of "poaching"—a term to describe how a person not hailed by a particular text will appropriate it and rework the text for

him or herself. Reading *The Jim Rome Show* through irony and sarcasm fits with Duncan and Brummett's (1993) notion of radical empowerment—a subversive subject position for reading sport texts. In
their study of spectator groups viewing televised football, they argue
that women viewers acted upon the text to empower themselves.
Some female viewers used a strategy called "liberal empowerment,"
which involved using televised football to increase their knowledge of
and identification with the game. This strategy has limits, according to
Duncan and Brummett, because it accepts the patriarchal premise on
which televised football rests. Radical empowerment occurred when
women spectators subverted the premises of the televised football
spectacle, using irony, sarcasm and a limited commitment to the game
and to the patriarchal readings of the text. In undercutting the seriousness of the football game, these female spectators played an active
role in their own agency, actively resisting an institution that privileges
male dominance. This tactic is in stark contrast to those of many male
viewers, who tended to empower themselves by adopting a subject position of the mainstream fan, by expressing strong identification with
the game and displaying knowledge, which allowed them to boast
about their sports expertise.

Joan's not taking Rome seriously (as opposed to some of the male
clones) and seeing the irony of the show were her ways of empowering
herself and not internalizing the patriarchal values and beliefs promoted by *The Jim Rome Show*. Joan's textual poaching can be an important tactic that marginalized groups of consumers might need to
take in order not to see themselves as victims. Other strategies Joan
used were direct protest by calling the show and by offering a different
viewpoint on the inherent sexism and homophobia in sports. In regard to sexuality, Joan appreciates Rome's liberal stance in supporting
gay male athletes who "come out," but she is critical of his view of lesbian athletes, particularly referring to Martina Navratilova as "Martin." "That infuriates me and I turn the dial. Until Rome honors
women and queer athletes, the show is still pretty conservative." Thus,
Joan uses the strategy of critically analyzing a text to challenge the patriarchal and heteronormative assumptions of sports radio.

MY TAKE ON THE AUDIENCE OF SPORTS TALK RADIO

My audience research allowed sports radio texts to become alive and
suggested that such texts are not necessarily deterministic and univocal.

Rather, my research suggests that sports radio texts are sites of struggle when audience members and their various subject positions are taken into consideration. While there were few signs of authentic resistance to the patriarchal discourses in sports radio (an exception being Joan), my interviews suggest that there are a number of contradictions and ambivalences whose cultural and gender significance should not be underestimated. Evidence of one such contradiction is found in several of my interviews with men who support Jim Rome's liberal stance on homosexuality and sports. It is through these contradictory and ambivalent spaces that sports radio may have its greatest political potential in terms of what it may signal for changing gender and sexuality relations and identities. Hence, my interviews confirm my view that sports radio should not be interpreted in purely negative terms as a backlash to feminism. Rather, sports radio is a commodified and mediated "homosocial" space for some men to negotiate and rework new versions of masculinity.

For some men, I suggest that some of the topics discussed on sports radio may play a role in their being able to discuss gender and sexuality issues in a forum that helps them develop new awareness in a publicly mediated way. It is safe, under the guise of "sports talk," for many fans to discuss larger, societal issues; there is an ethic of sports fandom that allows some discursive space for men to convey opinions in a passionate, even aggressive, but respectful manner. Since sports is embedded with so many sociopolitical issues, it is inevitable that larger social issues surface in sports conversations. Talking about sports, eating chicken wings, and drinking beer provide the anchor for many men to only not discuss politics but also to allow some latitude to "open up" and take some risks in supporting counterhegemonic discourses. My research illustrates that men tend to express their gender anxieties or support of gay athletes during one-to-one interactions as opposed to group contexts. These one-on-one encounters, particularly when the tape was turned off, gave some of my subjects the rare opportunity to discuss issues of sexuality and gender in a nonthreatening manner.

One area in which sports radio may have its greatest political potential is the opportunity the genre allows sports fans to have a voice and critique the power and hegemony of big-market teams and owners. Susan Faludi (1999) argues that postwar society has "stiffed" American men by making them a number of promises regarding their role as fathers, the workplace, the economy, and the political community that it has failed to deliver. Now the institution of sports has also failed American men, having been corrupted by the ethics of corporate capitalism; the virtue and romance of their favorite teams and players have been stolen from them due to money-hungry owners and disloyal players. Working-class men, in particular, cannot afford to attend games as their fathers have in the past. Moreover, their favorite players or teams leave for a more financially lucrative arrangement. Sports radio may be the only public mediated site where, for the small price of a telephone call, fans can express their outrage. This indignation could be utilized to educate sports fans about the parallels between the economics of major sports and the larger, global economy.

While acknowledging the civic potential of sports radio, we must realize that it is, however, a genre with several constraints, including its link to commercialism and its celebration of laddish form of masculinity. My fieldwork at the tour stops reveals that many consumers of sports radio project an unproblematic and undifferentiated image of natural masculinity that, while possibly ironic and stylized, is not without consequences, including the danger and sexual harassment that Debora experienced.

My research in sports bars and tour stops was both enjoyable and fraught with ethical dilemmas. At times, I was uncomfortable with the laddish displays of manhood and frequently feigned laughter. Admittedly, I also participated in crude, sexist jokes on occasion, keeping secret my feminist-informed values. I also was oblivious to Debora's being hassled by a clone. Yet, I simultaneously enjoyed my fieldwork experience and genuinely felt a part of this community. I know that a key reason why I feel I belonged in this community is my privilege as a white male sports fan.

How do I make sense of a feminist standpoint in my research while concurrently participating in and enjoying the sports radio community? Is it hypocritical of me to have participated, at least occasionally, in "male collusive discourse"? Whether it is duplicitous or not, I have come to realize that I cannot rise above the culture, despite having some cultural capital and training in masculinities scholarship. My personal experience in sports bars and tour stops mirrors contemporary masculinities—a set of conflicting, hybridized, and unstable discourses that range from hegemonic to pro-feminist, from new man to new lad. My behavior as an ethnographer unfolded along this continuum, similar to that of many men in American culture who are trying to make sense of manhood in a postfeminist world and the isolation created by the global economy. The sports radio community, whether in radio space or in the physical space of a sports bar, is a relatively inexpensive space in which men can break free of the isolation of commercial culture, build community, and rework ideals of manhood.

It is imperative that male scholars write from a more personal standpoint in regard to their contradictory experiences of being a man in a patriarchal culture rather than just from a privileged academic and theoretical position. My research into sports talk radio inspired me to become more aware of my privilege as a white, heterosexual man. I, like many of the men I interviewed, often feel isolated due to increased work demands and the disappearance of public space. Sports talk, although enclosed within corporate consumerism and patriarchy, partially fulfills my desire for a third place—a place in which to meet and interact with other fans.

9   9th Inning

MY FINAL TAKE

I usually had sports talk radio or televised sports in the background while writing this book. Listening to sports comforted me and justified having sports programs turned on as "part of my research." While I was writing this conclusion, I was listening to *The Best Damn Sports Show Period* (March 30, 2003) in which Chris Rose and actor Tom Arnold interviewed professional golfer Rosie Jones about her recent decision to publicly "come out" in a column she wrote in the *New York Times*. Jones (2004), a successful golfer on the LPGA, wrote in "First, a Word about Me and My Sponsor" that the impetus behind her public declaration was her association with a new sponsor, Olivia, a lesbian travel company. Jones admitted to Arnold and Rose that she was nervous about making the move because she feared criticism from sportswriters and fans and the loss of corporate sponsors Yet Rose and his sidekick Arnold congratulated her courage, stating that her coming out might be "one more step toward ending homophobia in sports." Rose also mentioned that Jones's sponsors have all decided to support her despite her publicly declaring her lesbian identity. "This is huge," Rose replied. "A decade ago Martina Navratilova lost her sponsors when she came out—this is real progress." I also thought to myself that this segment, on a very macho and misogynistic program, was disruptive of heterosexual dominance in sports. However, my hope was tempered in the show's subsequent segment, when Arnold and football player Eric Dickerson ridiculed Rose for wearing a "sissy shirt." In a matter of two quick segments, the show ranged from antihomophobic and gender-sensitive (interviewing Rosie Jones) to sexist (feminizing Chris Rose).

This example illustrates how sport interfaces with larger social issues, such as sexuality and gender, and demonstrates that sport and the sports/media, while historically viewed as anti-intellectual, are important sites for critical analysis. This study uses sports talk as a text to understand contemporary masculinity and its intersection with class, nation, sexuality, and race. Connell (2000) argues that sport is the leading definer of masculinity in Western culture; it serves as a key cultural site in which hegemonic masculinities are created, shaped, and performed, involving the exclusion of women and "other" (gay and nonwhite) men. The study of sport talk radio raises a number of questions concerning masculinity and its association with media and cultural power, democracy, and the influence of commercial culture. In this final chapter, I seek to clarify the contribution this book has made to these overlapping areas of analysis by focusing on three themes.

As noted in the introduction, this book seeks to articulate a complex position between some of the traditions of media and cultural studies, sports studies, and masculinity studies. By listening carefully to my research participants and by blending core concepts from relevant disciplines, I have staked out a middle position describing the temporal and fragmented state of contemporary masculinity. Hence, terms such as "ambivalence" and "ambiguity" have marked my position on sport talk radio rather than extreme celebratory or moralistic accounts of the genre. Rather than identifying with either extreme, my research has revealed a range of readings that are indicative of the ambivalences and instabilities of contemporary masculinities. These ambivalences, or what I referred to as "unique outcomes," demonstrate some potential for change in gender relations and identities. I will discuss three areas in which sports talk radio opens up avenues for transformation: (1) civic discourse;(2) sexuality; (3) masculinity.

SPORTS TALK AND CIVIC DISCOURSE

The chapter on the production of sports talk radio highlighted the explosion of mergers and ever-narrowing control over sports talk radio. The Telecommunications Act of 1996, in particular, removed previous obstacles to ownership of multiple stations in the same market, provoking a surge of station purchases and consolidations of territory. This act, along with the abandonment of the Fairness Doctrine, has weakened democracy and allowed right-wing politics to dominate the airwaves. Right-wing talk radio emerged in the early 1980s in the form of Rush Limbaugh, an articulate and talented talk show host out of Sacramento. By 1987, Rush rose to national status by offering his program free of charge to stations across the nation. Station managers, not being business dummies, laid off local talent and picked up Rush's free show, leading to a national phenomena: the Limbaugh show was one of America's greatest radio success stories, spreading from state to state faster than any modern talk show had ever done. (Such free or barter offerings are now standard in the industry.)

Station managers also discovered that there is a loyal group of radio listeners (around twenty million occasional listeners, with perhaps one to five million who consider themselves "dittoheads") who embraced Rush's brand of overtly hard-right spin, believing every word he says, even though he claims his show is "just entertainment" to avoid a reemergence of the Fairness Doctrine. The success of Rush led local radio station programmers to look for more of the same: there was a sudden demand for Rush-clone talkers who could meet the needs of the nation's Rush-bonded listeners. The all-right-wing-talk radio format emerged, dominated by Limbaugh and Limbaugh clones in both style and political viewpoint. Thus, the extreme fringe of the right wing dominates talk radio not because all radio listeners are right-wingers but because the right wingers and their investors were the first to market a consistent and predictable programming slant, making right-wing-talk the first large niche to mature in the newly emergent talk segment of the radio industry.

Sports talk radio is owned and operated by the same companies who operate political, right-wing talk radio. According to these hegemonic corporate entities, sports talk radio is a triumph of government deregulation and other radio industry changes—namely, "niche marketing" and "narrow-casting," which turned radio into an increasingly commercial space. My interviews with people in the sports radio industry note some of the journalistic compromises they make in order to please corporate sponsors. David Theo Goldberg (1998) suggests that sport radio reinforces the antidemocratic agenda of global capitalism by pushing new products to a guaranteed niche market of twenty-five to fifty-four-year-old white men. He argues that sports talk radio, like its cousin, political talk radio, reproduces a community of white, male, like-minded consumers that indicates the "death of civil discourse as social control through fanaticism takes over" (p. 221). Yet my study contradicts much of Goldberg's thesis, noting some of the exceptions to corporate and capitalist hegemony. My project is more in line with that of Haag (1996), who thinks there is a democratic force inherent in sports radio because it fashions civil talk in public space despite its being disseminated by media conglomerates.

Like Haag's argument that sports fandom allows fans to talk passionately about sports through civil disagreement, my analysis notes that sports talk has moments of inclusive and respectful dialogue. And while there are a range of in-studio regulatory strategies to control callers, many fans experience the radio airwaves as a rare space in which to "have a voice." Many of these men, feeling uncertain about gender and economic changes, find this airwave community to be somewhat egalitarian. As stated before, Oldenburg (1989) suggests that "third places" are not important solely on the psychological level; they can be important on a political level as well. Unfortunately, third spaces have been disappearing due to social and economic changes, including suburbanization and the private commercialization of space. While sports radio is distributed by private commercial forces that often reinforce male privilege, it does provide much of the value found to exist in third spaces. Sports radio does not necessarily replace the existence of physical public space and may even be a poor substitute that actually shrinks public space. Yet sports radio's popularity suggests male desire for connection, breaking out of the living conditions that contribute to isolation. Moreover, sports talk radio's appeal is a call to action to work in the face of private commercialization of space, to preserve existing third spaces, and to develop substantive ones that reinvigorate grass-roots civic society.

In addition, while sports talk radio is commercialized, it also provides rare opportunities for hosts and fans to critique global capitalism. Because of the success of corporate dominance in global and domestic affairs, the trickle-upward effects of current economic policies are seen as existing outside of politics and culture and therefore not subject to cultural critique. Sports talk radio may be one media site in which it is permissible to criticize corporate culture rhetoric; the breadth and depth of critical comment in sports radio stand in stark contrast to "mainstream" news. While the mainstream press handled the Enron debacle as a mere aberration, for example, sports talk radio has been largely critical of systematic capitalist indiscretions in the sports world—whether exorbitant free agent signings or the increasing number of franchises moving to another city looking for greener pastures. Sports radio allows sports fans to vent their anger. It is difficult to imagine a newspaper or television news channel devoting a segment to the inequities produced through global capitalism, but often sports talk radio and other sports media forums seem altogether concerned with disparaging the finances and corporate structure of professional sports. So despite sports talk radio being distributed by powerful corporations, the talk produced on the airwaves is not completely colonized through consumer capitalism. Indeed, sports talk radio creates a potentially productive location in which to educate people about larger inequities created by consumerism and corporate hegemony. In summary, sports talk radio serves as a potentially democratic and oppositional space in some regards, particularly in the area of unmasking the ideology of corporate capitalism.

SPORTS AND SEXUALITY

Research on sports radio (Goldberg, 1998; Haag, 1996; Mariscal, 1999; Smith, 2002) suggests that dominant ideologies of gender, race, sexuality, class, and nationalism are perpetuated in its coverage of sports. For instance, the dominant discourses around gender and sexuality are reproduced through trivializing women's sports and celebrating traditional masculinity. This traditional gender logic is often reinforced by homophobic speech, with the fear of lesbianism lurking under the surface of much sports talk. Yet my study notes the significantly antihomophobic tenor of *The Jim Rome Show*, which may provide an opportunity to address heterosexual dominance in sports. Yes, Rome's radio program reveals the limits of liberal ideology, but his progressive stance may be a starting point for influencing heterosexual men.

These progressive moments in sports talk radio are significant, since Rome and other sports talk hosts likely have more influence on the typical heterosexual sports fan than do academic scholars and political activists. The participants I interviewed, while generally holding conservative opinions about gender and sexuality, were influenced by Jim Rome's antihomophobic posture and support a tolerant perspective. These progressive moments in sports talk radio can be used by scholars and gay, lesbian, bisexual, and transgender (GLBT) activists to advocate tolerance, provide support for gay and lesbian athletes, and challenge institutional heterosexism in sport. Further research should explore whether these fissures in sports talk texts can extensively transform the beliefs of men who have not yet acquired a sincere commitment to antihomophobia.

While acknowledging Jim Rome's liberal stance on homophobia, it is important to mention its limitations. Rome's tolerance of gay athletes is firmly situated within the new homonormativity (Duggan,

2001) that is at odds with a more progressive political agenda. The homonormativity that Rome sanctions on his show is an assimilationist, nonpolitical, and consumerist version of sexuality that advocates only the right of gays to be in the military and married, claiming just a small private domestic space for gays in the neoliberal world order. This homonormativity redefines key terms of traditional gay politics, and the right to privacy becomes domestic confinement in a corporate culture managed by a minimalist state. What's more, this assimilationist, conservative view of sexuality tends to privilege the white middle-class male as the universal gay subject. Rome and other sports talk hosts poise praise for virile gay men such as 9/11 hero Mark Bingham, the gay rugby player, or Billy Bean, the ex-baseball player, while making invisible or mocking lesbian athletes such as Martina Navratilova.

The new homonormativity is quite evident in programs such as *Queer Eye for the Straight Guy*, in which heterosexuality and consumerism are at the center and in which homosexuals will be accepted and cherished, as long as they do not claim any real power to change these structures. The neoliberal drift toward a consumerist, depoliticized gay/lesbian sexuality is even evident with recent changes in the organization Gay Lesbian Alliance against Defamation (GLAAD), a GLBT media advocacy group that funded part of my research through its Center for the Study of Media and Society (the only research center in the United States to exclusively fund and distribute studies that examine the ways in which news and entertainment media represent LGBT people and events). While writing this chapter, I received an email from the director of the Center for the Study of Media and Society saying that GLAAD has made a decision to move away from funding academic research. In its place, GLAAD will be commissioning marketing research and large-scale polling studies outsourced to the private sector.

Sports talk radio, even with its liberal, antihomophobic stance on shows such as Rome's, tends toward a conservative, neoliberal "reproductive agency" (identification with corporate consumerism that stabilizes oppressive social institutions) than a more radical or transgressive "resistant agency" (Dworkin and Messner, 1999). Hence, while the "third space" of sports radio may be welcoming of gay males, it is not necessarily a progressive sexual space, since it prevents a substantive discussion of the structures of heterosexual domination that excludes lesbians and other bodies that do not fit the gender norms of our culture. According to Oldenburg (1989), the best third places are those that are inclusive (which sports radio is not). As I have argued, while the mediated community of sports talk radio does produce male camaraderie, it is a fairly segregated space.

Clearly, sports continue to reinforce homophobia, but it is undergoing immense change, with sexuality at the center. Sure, sports radio, with its link to commercial capitalism, is not a site of revolutionary sexuality. Nevertheless, sports talk radio programs such as *The Jim Rome Show* have an ambiguous but distinct investment in the topic of sexuality. This ambiguity indicates a shift in sexual and gender relations that is a sign, and possibly a source, of social change. One area in which this change is occurring is the way lesbians, such as Joan, listen to sports radio texts. Due to the threat of lesbianism in sports, women who perform female masculinity by calling sports radio or playing traditional male sports, such as football or rugby, challenge the traditional gender and sexual logic. According to Miller (2001b), a great deal of lesbian writing has focused on sports as a venue for meeting people, forming alliances, and creating community. When lesbians such as Joan intrude on the male preserve of sports radio, they force men to reexamine the gender order.

MASCULINITY AND SPORTS

Michael Messner (2002) writes that at "a historical moment when hegemonic masculinity has been destabilized by socioeconomic change and by women's and gay liberation movements, the televised sports manhood formula [sports media] provides a remarkably stable and concrete view of masculinity as grounded in bravery, risk taking, violence, bodily strength, and heterosexuality" (p. 126). My research notes that much of sports talk radio, similar to televised sport, largely reproduces and naturalizes existing gender, racial, sexual, and class hierarchies. In addition to many textual examples that confirm these inequalities, in my research at tour stops I observed laddish displays of masculinity based on homophobia, sex and alcohol, self-centeredness, and the exclusion of women: practices promoted by the values and norms of sports talk radio. Surely sports and, by extension, the sports media confirm that hegemonic masculinity is firmly in place.

Yet hegemonic masculinity is not a static phenomenon, but is always shifting and open to change. This book notes that many of the men I interviewed exhibited a broad range of seemingly inconsistent and ambivalent behaviors ranging from the misogynistic to the egalitarian.

McKay, Messner, and Sabo (2000) suggest that there is a tendency for sport sociologists and masculinity scholars to overemphasize negative outcomes for men in sports. They argue that this overemphasis on negativity leads to simple conclusions that sports is only sexist and homophobic. McKay, Messner, and Sabo encourage more nuanced analysis of men's sports experiences to examine the extent to which, even within conventional and institutionalized athletic contexts, there is room for gender display or even resistance to hegemonic masculinity. Laurence de Garis's (2000) analysis of the intimacy that develops between male boxers and Alan Klein's (2000) research on the vulnerability expressed by Mexican baseball players are two such studies that challenge dominant assumptions about the "shallowness" that often results from male athletes' competitiveness, aggression, and homophobia. This book builds on this research by highlighting some of the contradictions in and disruptions of hegemonic masculinity. My intention in interpreting sports talk radio in more ambivalent terms is based, in part, on a desire to move beyond masculinity studies' tendency to focus only on the destructive features of traditional manhood and the "costs and benefits of male privilege." Despite the grim outlook for masculinity described in much academic work, the fact remains that we still need to know how men perceive masculinity, whether they experience masculinity-in-crisis, how they enact "masculinities," and how they bond with other men and women. My project, by using sports radio texts and audiences as the vehicle, attempted to carry this out.

Surprisingly, some of the interruptions of hegemonic masculinity occurred at the sports bars I frequented over the past three years. I have generally enjoyed the experience and, over time, felt part of a "real" community. The actual physical public space of a sports bar allows the imagined sports radio community to be more "real" through face-to-face interaction. Although listening to and calling sports talk radio shows may offer some possibility for playful experimentation with gender identity, it cannot provide all the values attributed to in-person interaction (such as feelings of good will, physical contact, and more personal friendship). This third place community, I argue, does serve some positive functions, including reducing isolation and creating new relationships among men (and women) of similar interests—in this case, sports. While acknowledging potential gender blind spots, I did experience many moments of connection with other sports fans, although often in the form of "covert intimacy" (Messner, 1992).

Even the more overt displays of homoerotic, gender-bending relations—the butt slapping and rowdy hugs—reminded me of moments in boyhood when a pre-adult, pre-adolescent joy of intimately touching had not been defined or theorized as potentially progressive. Other moments of intimacy occurred in one-to-one encounters when many men shared a more egalitarian, gender-sensitive side of their personhood. I am not suggesting that the egalitarian discourses and practices at the bars disrupt or contribute to broad relations of male dominance. Although certain egalitarian practices may be related to sexual and gender equality, there is no real substantial evidence in my study that the men I interviewed reached a zenith of egalitarian insight that extended to other areas of their lives. However, it was significant in my eyes and suggests that men's interactions in "third spaces" are not simply places of feminist backlash. It is in that sense, I would posit, that sports talk radio signifies the possibility that alternative masculinities might emerge even as sports radio reinscribes traditional forms of manhood.

In summary, sports talk radio and sports fandom are full of weird contradictions. Sports fandom is a significant part of many men's (and increasingly women's) lives, and despite all the "academic babble" the subject attracts, it provides many people an opportunity to briefly bolster the quality of their emotional and social lives. For the fans of sports talk radio, there are fleeting moments of joy that can, albeit temporarily, escape neoliberalism and hegemonic masculinity. There is delight in watching one's favorite player hit a game-wining home run; there is genuine closeness when talking about your home team with a stranger; there is real passion in debating whether Lebron James or Dwayne Wade is better basketball player; there is even satisfaction in laughing at one of Jim Rome's jokes while driving to work. There is pleasure when I remember the '68 Tigers winning the World Series. For these moments, sports talk radio is lifted out of the mire of corporate sponsorship, advertising, and patriarchy and provides flashes of pleasure that are enlivening. That is my last take. As Jim Rome says at the end of every show, "I'm out."

NOTES

1. OPENING PITCH: THINKING ABOUT SPORTS TALK RADIO

1. According to Arbitron ratings, *The Jim Rome Show* is ranked eight in radio talk audience share. The most popular radio talk hosts, according to the ratings, are Rush Limbaugh and Dr. Laura. All three shows, Jim Rome, Rush Limbaugh, and Dr. Laura, are owned by Premier Radio Networks, a company worth 330 million dollars. These statistics are from Premiere Radio's Web site—www.premiereradio.com.

2. "Run" refers to the host hanging up on the caller.

3. Jim Rome's Web site (jimrome.com) has a twenty-four-page glossary (known as "city jungle gloss") that lists his terms and the definitions. For instance, "jungle dweller" refers to a frequent telephone contributor to Rome's show. "Bang" means to answer phone calls. "Bugeater" refers to a Nebraskan who is a fan of the Nebraska Cornhuskers college football team.

4. The comment "What will Rome say next?" has been applied several times to listeners of the Howard Stern show, both for those who enjoy and despise it. This is (even) mentioned in the Howard Stern autobiographical movie, *Private Parts*.

5. Deregulation was championed by then FCC chairman Mark Fowler, who sold it as a form of media populism and civic participation. However, this public marketing campaign masked increased economic consolidation and increased barriers to entry into this market for all but very powerful media conglomerates such as Infinity Broadcasting and Premiere Radio. Commenting about the success of conservative white male talk radio due to deregulation of the 1980s, Susan Douglas (2002) claims that Reaganism was successful by "selling the increased concentration of wealth as a move back toward democracy" (p. 491).

6. In 1960, there were just two radio stations in the United States that were dedicated to talk radio formats (Goldberg, 1998).

2. THE SPORTS TALK RADIO INDUSTRY: FROM RUSH TO ROME

1. Alan Berg's story became the vehicle for a movie directed by Oliver Stone titled *Talk Radio* (1988).

2. There is a growing liberal alternative to right-wing radio such as "Air America" and the "Ed Schultz Show."

3. These figures are from Premiere Radio's Web site, http://www.premrad.com.

4. These figures are derived from Arbitron radio's Web site, www.arbitron.com. The figures I cite here are from a power point presentation titled "Handicapping the Ratings: Inside the Numbers of America's Sports Stations" by John Snyder (2002).

3. INSIDE THE SPORTS RADIO INDUSTRY: ADS AND LADS

1. I attempted several times via phone and email to contact *The Jim Rome Show* to interview Rome or his production staff. My attempts were unsuccessful. This experience highlights some of the difficulties of access in ethnographic research.

2. In Sacramento, one of these important stories is a plan for an expensive downtown sports arena. The Kings, the City of Sacramento, and Union Pacific jointly have put up $800,000 to study the feasibility of a basketball facility in downtown Sacramento. It is likely that the owners of the Kings will ask taxpayers for hundreds of million dollars to build this new arena. Covering that issue could put media outlets that are tied to the Kings (such as KHTK) in a compromising position.

3. By neoliberal ideology, I am referring to the privileging of market forces and the dismantling of the welfare state.

4. Newton (2005) argues that the Promise Keepers, like sports talk, can also break down racial barriers.

5. In my research, I learned that currently only four women are working as hosts in the sports talk industry. Only one woman, Nanci Donnellan, has ever hosted a national show. Although quite popular, and a gender rebel, she was unexpectedly fired by her network (Sports Fan Radio Network) in 2001. Some sports journalists have argued that Nanci Donnellan was fired as Rome's popularity increased.

6. Also, her comments describe a common habit of many men who "open up" and express part of themselves to women when no other men are in their presence.

4. *THE JIM ROME SHOW*: "MYSPACE.COM" FOR MEN

1. "Skank" refers to a woman who is promiscuous.

2. *Queer Eye for the Straight Guy* is a popular television show on the cable channel Bravo. The show stars five gay men who do a home fashion makeover with a straight man each week. Each week their mission is to transform a style-deficient and culture-deprived straight man from drab to "fab" in each of their respective categories: fashion, food and wine, interior design, grooming, and culture.

3. This is different from local sports talk radio, where there is much more dialogue and back-and-forth interchange between the callers and hosts.

4. Rome's relationship with his caller, like most talk show power relations between caller and host, is quite asymmetrical. Ian Hutchby (1996), in his study of the discourse in talk radio, states that although the host has an array of discursive and institutional strategies available to keep the upper hand, occasionally callers have some resources to resist the host's powerful strategies. Hence, Hutchby argues that power is not a monolithic feature of talk radio. Hutchby's argument does not appear to work with *The Jim Rome Show* as callers hardly ever confront Rome's authority. Rather, Rome's callers want his approval.

5. Vince Foster was President Bill Clinton's White House Counsel who apparently killed himself in 1993. After his death, several reports questioned whether Foster actually committed suicide, claiming that the forensic evidence pointed to a murder. Many conspiracy theories arose stating that Foster was murdered because he had personal evidence that Bill Clinton and some of his staff engaged in illegal activities, including financial improprieties.

6. Tremblay and Tremblay (2001) noted that 17 percent of the calls they studied during a two-week period of *The Jim Rome Show* were run.

5. RACE, GENDER, AND SEXUALITY IN THE JUNGLE

1. Belinda Wheaton (2000) suggests that alternative sporting cultures are potential sites for more progressive sporting masculinities.

2. In 2004, steroid use among professional baseball players was a big media spectacle. On February 11, 2004, San Francisco Giant Barry Bonds's personal trainer Greg Anderson had his house raided in September and the substances found in his possession are suspected to be steroids. They also found materials that detailed which athletes were on which drugs and their respective intake schedules. In his 2004 State of the Union speech, President George W. Bush condemned the rampant use of steroids by professional athletes and called upon players, coaches, and owners to eliminate the use of performance-enhancing drugs. Gary Whannel (2002) states that drug

testing has less to do with new concerns about the health risks of using drugs but more about the "new corporate paternalism, whereby institutions become the new moral guardians of their employees, supervising the way they live" (p. 157).

3. However, it is an important to note that Rome asserts his authority over a person with less power—a first-time caller. Rome doesn't take this strong a stance with Eric Davis, a high-status person who likely has more influence within the sports world. This textual example reveals the power relations of talk radio; hosts and famous athletes have more authority than callers.

4. Both Goldberg (1998) and Mariscal (1999) suggest that sports talk radio is more racialized than any other radio format.

5. Mariscal (1999), in his analysis of *The Jim Rome Show,* notes Rome's contradictory stance on race, stating that at times Rome is very progressive and antiracist, and other times Mariscal notes that Rome engaged in derogatory stereotypes toward Latinos. Mariscal states that Rome's inconsistent stances on "racially charged topics reveal the basic slippage in liberal discourse" (p. 116), a situation where citizens engage in post-civil rights speech that "slides easily from tepid antiracism to the reproduction of deeply ingrained racist clichés" (p. 116).

6. Mariscal (1999) posits that Jim Rome's liberal show moved to the right after national syndication, reflecting the ideology that dominates other sports talk radio programs

7. I teach a graduate seminar in the Division of Social Work at California State University, Sacramento, titled "Diversity and Multi-Cultural Practice." The class engages critically with race, gender, sexuality, ability, and class and includes an examination into the critical study of whiteness and white privilege.

8. Hamilton was forced to resign his position as a talk radio host after describing Japanese baseball pitcher Hideki Irabu as a "fat Jap." He also said that African American football player Lawrence Phillips needed to be "lynched."

9. Capitalized words suggest loud talk.

10. Even listeners who do not call can feel involved in the process due to their active listening.

11. Also, note Rome and the caller's appropriation of African American vernacular discourse in their interchange.

12. By whiteness I am referring to various critical scholars who view whiteness as a social construction rather than a "natural category" (Frankenberg, 1997; Nakayama and Krizek, 1995). A social constructionist view of whiteness notes that its meanings are produced by "socially and historically contingent processes of racialization, constituted through and embodied in a wide variety of discourses and practices" (Wray and Newitz, 1997, p. 3). Critical whiteness scholars attempt to interrogate the discursive and social power of whiteness by the ways it makes itself invisible, eluding analysis yet

exerting influence and dominance over everyday life. It is important to also note that whiteness is not monolithic and fixed but rather conflicting and contradictory. Whiteness always is interacting with gender, class, nation, and sexuality.

13. Maureen Smith (2002), in her content analysis of Rome, noted that Rome engaged in racialized comments by associating several black athletes who engaged in violent behavior with O. J. Simpson. For instance, Rome referred to Carolina Panther Rae Carruth as "Raenthal" after he was found guilty of murdering his wife. During this same time period, Green Bay Packer Mark Chmura, a white athlete, was charged with having sex with a minor, which included his sitting in a hot tub at a prom party. His nickname was "American Chewy," with no linkage between Simpson and Chmura.

14. Allegations that Bonds used steroids were highlighted in a well-publicized book, *Game of Shadows*, by *San Francisco Chronicle* writers Mark Fainaura-Wada and Lance Williams (2006).

15. Smith (2002) notes that when Iverson produced a rap album prior to the 2000 NBA season, Rome was critical of Iverson's lyrics, telling his listeners: "Violence against women and gays is not OK." Numerous callers expressed their dislike for Iverson, not necessarily because they were opposed to his music, but because he was an athlete they did not like. This incident presented them with an opportunity to safely disparage Iverson without seeming to be personal. The Iverson rap CD was an issue that remained a topic for most of the remainder of October. Rome invited sports sociologist Harry Edwards on the show to offer his analysis. When Rome pressed Edwards for an explanation of why Iverson could not rap about positive images, such as his peers Kobe Bryant and Shaquille O'Neal, Edwards was quick to point out that Iverson had experienced a violent and oppressive upbringing and was simply writing about what he knew.

16. The link between nationalism, sport, and the Iraq/Afghanistan War was evident once again with death of ex-NFL player Pat Tillman. Tillman was a defensive back for the Arizona Cardinals who decided to leave the NFL (and his salary) to join the Army Rangers. Tillman said he was deeply affected by 9/11 and felt he needed to do something for his country. He was killed in action in April 2004 in Afghanistan. Tillman's decision to give up his lucrative salary and NFL celebrity status to "fight for freedom" was viewed by Rome and other sports media figures as the ultimate sacrifice. Rome referred to him as a "true American hero" and a "real patriot." Rome was asked by Tillman's family to be the master of ceremonies at the funeral service. Rome said that being the master of ceremonies was "the greatest honor of his professional life" (*The Jim Rome Show*, May 4, 2004).

17. United States exceptionalism was evident in the 2006 World Cup Soccer tournament. Rome consistently derided the tournament, the game of soccer, and the rest of the world for loving soccer.

18. During the halftime of the 2004 Super Bowl football game, popular music stars Janet Jackson and Justin Timberlake were performing as a duo. During their performance, Timberlake reached over and tore Jackson's top, exposing her breast. The FCC called for a "thorough and swift" probe of Janet's exposed breast. A five-second delay for censors was instituted at the Grammys and the Oscars.

6. IN THE JUNGLE WITH THE "CLONES"

1. It is also important to note that many audience response approaches examine the real-time consumption of a text. My approach to reception puts some distance of time and context between the inquiry and the act of listening to sports radio.

2. I invited several women to be interviewed about *The Jim Rome Show*. However, only two stated that they listened to the show.

3. The local Sacramento sports talk affiliate runs a commercial that says, "Belly up to the bar and pour yourself a cold one! You are listening to your sports bar on the radio."

4. There is an ongoing debate in cultural studies about postmodern condition. The debate centers on whether postmodernism allows for diversity and appreciation of difference or whether the postmodern condition works in the service of global capitalism.

5. Messner (1992) defines "covert intimacy" as doing things together rather than mutual talk about inner lives.

6. Postfeminism refers to the idea that women have already achieved full equality with men (Humm, 1995).

7. WHERE EVERYBODY KNOWS YOUR NAME

1. There is a long history of sports and illegal gambling, including the 1919 "Black Sox" scandal in which several members of the American League Pennant winner Chicago White Sox deliberately lost ("threw") some games in order to be paid by bookies. Also, popular baseball player Pete Rose has been "banned" from baseball due to illegally betting on baseball while he was managing the Cincinnati Reds.

2. This information was given to me by KHTK's station manager.

3. My experience at the tour stop also illustrates the ethics and complications of doing ethnography in hegemonic spaces.

4. My survey of the fifty clones we talked to in Detroit Tour Stop parking lot found out that thirty-five were single, eleven were married, and four were divorced. The jobs ranged from mechanic, truck driver, auto worker, teacher, attorney, family therapist, student, engineer, and sales. The two

most held jobs were drivers and sales. The two women we interviewed stated they were there because their husbands "made me." One woman was taught Rome's smack language on the way to the tour stop.

5. During Jim Rome's twentieth tour stop in Sacramento in 2000, several of Maureen Smith's (2002) undergraduate students, armed with clipboards and surveys, headed into the Arco Arena parking lot and surveyed over 200 attendees of an estimated crowd of more than 16,000. The survey was short and sought only to find out who the attendees were demographically (age, marital status, and job) and as Rome followers (hours listened to a week, favorite topics, and why they were at the tour stop). The 208 men (126 single, 76 married, and 6 divorced) surveyed were from a variety of jobs, including photojournalist, geologist, broker, school psychologist, banker, electrician, coach, carpenter, student, machine operator, attorney, small business owner, landscaper, construction, meter reader, auto tech, supervisor CYA, teacher, graphic artist, air quality planner, farmer, engineer, family therapist, cabinet maker, student, military, merchandising, grocery clerk, and accountant. The two most held jobs were those in sales and drivers (delivery and UPS). Favorite topics listed by the respondents were Kings, Orenthal, Neckar, pro sports, smackoff, and Rayanthal. There were a number of reasons the men attended the tour stop: hang out with friends without wife, Love Rome, he is God, drink beer, break from wife, Rome is God, drink beer and get away from wife, party, sports fan and true clone, see the pimp/need karma (sic), hang out with my friends and drink beer, to see Rome and be a part of the mother of all tour stops, To see the "King of Smack," beer, away from wife, party with friends, because Rome is the pimp in the box, to get screwed up, party with the clones, get drunk and laid, to get drunk and listen to smack, I love Rome, curious, fun, new experience, see the Kings. While there were fewer females in attendance, 27 (14 single, 12 married, 1 divorced), they were given the same survey as the men, with some varied responses. They held jobs as a tax auditor, telecommunications, painter, insurance specialist, housekeeper, nanny, sales, UPS, computer programmer, and lab technician. The number one job among the women was as a nurse. OJ, smack, interviews, baseball, Rocker were listed as their favorite topics. When asked why they were attending, their responses were: to see Rome and Kings, Jim Rome, to see if Jim Rome's ear is fake, my husband really wanted to!, my husband made me, to see what kind of real response he gets, my boyfriend got free tickets, I love Jim Rome as much as Jim Rome loves Jim Rome, my sister, because sweetie got tickets, eat, drink and be merry (and my boyfriend made me), big fan of Rome, unequal ratio of men to women.

8. A SPORTS RADIO INTRUDER

1. Wheatley (1994), in her study of women rugby players in Europe, argues that these women athletes, like American female football players, by playing a sport associated with maleness, challenge heteronormativity and poke crevices in the male preserve of sport.

2. Title IX of the Educational Amendments of 1972 is the landmark legislation that bans sex discrimination in schools, whether in academics or athletics.

BIBLIOGRAPHY

Adorno, Theodor, and Max Horkheimer. "The Culture Industry: Enlightenment as Mass Deception." In *Mass Communication and Society*, edited by James Curran, 5–26. London: Sage Publications, 1997.

Allison, Lincoln. "Sport and Nationalism." In *Handbook of Sport Studies*, edited by Jay Coakley and Eric Dunning, 344–55. Thousand Oaks, CA: Sage Publications, 2002.

Anderson, Eric. *In the Game: Gay Athletes and the Cult of Masculinity*. Albany: State University of New York Press, 2005.

Andrews, David L. "The Facts of Michael Jordan's Blackness: Excavating a Floating Racial Signifier." *Journal of Sport & Social Issues* 13, no. 2 (1996): 125–58.

Ang, Ien. *Living Room Wars: Rethinking Media Audiences for a Postmodern World*. New York: Routledge, 1996.

Armstrong, Cameron B., and Alan M. Rubin. "Talk Radio as Interpersonal Communication." *Journal of Communication* 39, no. 2 (1989): 84–93.

Armstrong, Gary. *Football Hooligans: Knowing the Score*. Oxford: Berg, 1998.

Beck, Ulrich. *The Reinvention of Politics: Rethinking Modernity in the Global Social Order*. Cambridge: Polity, 1997.

Beynon, John. *Masculinities and Culture*. Philadelphia: Open University Press, 2002.

Birrell, Susan. "Discourses on the Gender/Sport Relationship: From Women in Sport to Gender Relations." *Exercise and Sport Science Reviews* 16 (1998): 459–502.

Birrell, Susan, and Mary G. McDonald. "Reading Sport, Articulating Power Lines." In *Reading Sport: Critical Essays on Power and Representation*, edited by Susan Birrell and Mary G. McDonald, 3–13. Boston: Northeastern University Press, 2000.

Birrell, Susan, and Nancy Theberge. "Ideological Control of Women in Sport." In *Women and Sport: Interdisciplinary Perspectives*, edited by D. Margaret Costa and Sharon R. Guthrie, 341–60. Champaign, IL: Human Kinetics, 1994.

Bolin, Anne, and Jane Granskog, eds. *Athletic Intruders: Ethnographic Research on Women, Culture, and Exercise.* Albany: State University of New York Press, 2003.

Boswell, Alan A., and Joan Z. Spade. "Fraternities and Collegiate Rape Culture: Why are Some Fraternities More Dangerous Places for Women?" In *Men's Lives,* edited by Michael Kimmel and Michael A. Messner, 167–177. Needham Heights, MA: Allyn & Bacon, 2001.

Bourdieu, Pierre. *Distinction.* London: Routledge, 1989.

Brod, Harry. *The Making of Masculinities.* Boston: Unwin Hyman, 1987.

Burns, Tom F. "Getting Rowdy with the Boys." *Journal of Drug Issues* 10 (1980): 273–86.

Burstyn, Varda. *The Rites of Men: Manhood, Politics, and the Culture of Sport.* Toronto: University of Toronto Press, 1999.

Butler, Judith. *Gender Trouble: Feminism and the Subversion of Identity.* New York: Routledge, 1990.

Buzinski, Jim. "Give the Media Good Marks: Coverage of Closeted Gay Baseball Player Was Positive and Non-Judgmental." 2001, www.outsports.com/columns/mediaoutmag.htm.

Carbado, Devon W., ed. *Black Men on Race, Gender and Sexuality.* New York: New York University Press, 1999.

Card, Claudia. *Lesbian Choices.* New York: Columbia University Press, 1995.

Carrington, Ben. "Race, Representation and the Sporting Body." In *CUCR Paper Series.* University of Brighton, 2002.

Cashmore, Ellis. *Sports Culture: An A–Z Guide.* New York: Routledge, 2000.

Caughey, John L. *Imaginary Social Worlds: A Cultural Approach.* Lincoln: University of Nebraska Press, 1994.

Clarke, Larry. "Sun Belt, Politicians Vie for Nascar Dads." *Washington Post,* August 2, 2003, A1.

Clatterbaugh, Kenneth. *Contemporary Perspectives on Masculinity.* Boulder: Westview Press, 1997.

Cloud, Dana. *Control and Consolation in American Culture and Politics: Rhetoric of Therapy.* Thousand Oaks, CA: Sage Publications, 1997

Coakley, Jay. *Sports in Society: Issues and Controversies,* 8th ed. Boston: McGraw-Hill, 2004.

Cole, C. L."Resisting the Canon: Feminist Cultural Studies, Sport, and Technologies of the Body." *Journal of Sport & Social Issues* 17 (1993): 77–97.

———. "Celebrity Feminism Nike Style: Post-Fordism, Transcendence, and Consumer Power." *Sociology of Sport Journal* 12 (1999): 347–69.

———. "PR Malfunction?" *Journal of Sport & Social Issues* 28, no. 2 (2004): 91–92.

Cole, C. L., and David L Andrews. "Look—It's NBA Showtime! Vision of Race in the Popular Imaginar." *Cultural Studies* 11 (1996): 141–81.

Cole, C. L., and David L. Andrews. "America's New Son: Tiger Woods and

America's Multiculturalism." In *Sports Stars: The Cultural Politics of Sporting Celebrity*, edited by David L Andrews and Stephen J Jackson, 70–86. New York: Routledge, 2001.

Collie, Ashley J. "Rome Rants." *The American Way*, August 8, 2001, 50–54, 56–57.

Connell, Ronald W. "An Iron Man: The Body and Some Contradictions of Hegemonic Masculinity." In *Sport, Men, and the Gender Order*, edited by Michael A. Messner and Donald F. Sabo, 83–95. Champaign, IL: Human Kinetics, 1990.

———. *Masculinities*. Berkeley: University of California Press, 1995.

Connell, Ronald W. *The Men and the Boys*. Berkeley: University of California Press, 2000.

Cook, Jackie. "Dangerously Radioactive: The Plural Vocalities of Radio Talk." In *Culture and Text: Discourse and Methodology in Social Research and Cultural Studies,* edited by Alison Lee and Cate Poynton, 59–80. New York: Rowman & Littlefield, 2001.

Cook, Kevin. "Media." *Playboy*, April 1993, 20–21.

Cornwall, Andrea, and Nancy Lindisfarne, eds. *Dislocating Masculinity: Comparative Ethnographies*. New York: Routledge, 1994.

Cosgrave, Jim, and Thomas R Klassen. "Gambling against the State: The State and the Legitimation of Gambling." *Current Sociology* 49, no. 5 (2001): 1–15.

Cox, Ana Marie, Freya Johnson, Annalee Newitz, and Jillian Sandell. "Masculinity without Men: Women Reconciling Feminism and Male-Identification." In *Third Wave Agenda: Being Feminist, Doing Feminism,* edited by Leslie Heywood and Jennifer Drake, 178–206. Minneapolis: University of Minnesota Press, 1997.

Crabb, Peter B., and Jeffrey H. Goldstein. "The Social Psychology of Watching Sports." In *Responding to the Screen: Perception and Reaction Processes,* edited by Byrant Bryant and Dolf Zillman, 355–71. Hillsdale, NJ: Lawrence Erlbaum, 1991.

Craig, Stephen, ed. *Men, Masculinity, and the Media*. Thousand Oaks, CA: Sage, 1992.

Cramer, Judith A. "Conversations with Women Sports Journalists." In *Women, Media, and Sport: Challenging Gender Values*, edited by Pamela J Creedon, 108–58. Thousand Oaks, CA: Sage 1994.

Creedon, Pamela J. "Women, Sport, and Media Institutions: Issues in Sports Journalism and Marketing." In *Mediasport*, edited by Lawrence A. Wenner, 88–99. New York: Routledge, 1998.

Creedon, Pamela J. "From Whalebone to Spandex: A Look at Women in American Sports Journalism." In *Women, Media, and Sport: Challenging Gender Values,* edited by Pamela J Creedon, 108–58. Thousand Oaks, CA: Sage, 1994.

Curry, Tim J. "Fraternal bonding in the locker room: A Profeminist Analysis of Talk About Competition and Women." *Sociology of Sport Journal* 8 (1991): 119–135.

Davis, Laurel R. *Hegemonic Masculinity in Sports Illustrated.* Albany: State University of New York Press, 1997.

de Certeau, Michel. *The Practice of Everyday Life.* Translated by Steve Rendall. Berkeley: University of California Press, 2002.

de Garis, Laurence. "Be a Buddy to Your Buddy: Male Identity, Aggression, and Intimacy in a Boxing Gym." In *Masculinities, Gender Relations, and Sport,* edited by Jim McKay, Michael A. Messner, and Donald F. Sabo, 87–107. Thousand Oaks, CA: Sage, 2000.

Douglas, Susan J. "Letting the Boys Be Boys: Talk Radio, Male Hysteria, and Political Discourse in the 1980s." In *Radio Reader: Essays in the Cultural History of Radio,* edited by Michele Hilmes and Jason Loviglio, 485–504. New York: Routledge, 2002.

Douglas, Susan J. *Listening In: Radio and the American Imagination.* New York: Random House, 1999.

du Lac, Jon F. "Frat Boy Nation: A New Culture of Chauvinism Buries the 'Sensitive Guy'." *Sacramento Bee,* 2002, EI, E7.

Duggan, Lisa. "The New Homonormativity: The Sexual Politics of Neoliberalism." In *Materializing Democracy: Toward a Revitalized Cultural Politics,* edited by Russ Castronovo and Dana D. Nelson, 175–94. Durham, NC: Duke University Press, 2002.

Duncan, Margaret C., and Barry Brummett. "Liberal and Radical Sources of Female Empowerment in Sport Media." *Sociology of Sport Journal* 10 (1993): 57–72.

Duneier, Mitchell. *Slim's Table: Race, Respectability, and Masculinity.* Chicago: University of Chicago Press, 1992.

Dunning, Eric. *Sport Matters.* New York: Routledge, 2002.

———. "Sports as a Male Preserve: Notes on the Social Sources of Masculinity and Its Transformations." *Theory, Culture and Society* 3 (1986): 79–90.

Dworkin, Shari L., and Michael A. Messner. "Just Do What? Sports, Bodies, Gender." In *Re-Visioning Gender,* edited by Judith Lorber, Myra Max Ferree, and Beth Hess, 341–64. Thousand Oaks, CA: Sage, 1999.

Eastman, Susan Taylor, and Arthur M. Land. "The Best of Both Worlds: Sports Fans Find Good Seats at the Bar." *Journal of Sport & Social Issues* 21, no. 2 (1997): 156–78.

Ehrenreich, Barbara. *The Hearts of Men: American Dreams and the Flight from Commitment.* New York: Doubleday, 1983.

Eisenstock, Alan. *Sports Talk: A Journey Inside the World of Sports Talk Radio.* New York: Pocket Books, 2001.

Faludi, Susan. *Stiffed: The Betrayal of the Modern Man.* New York: Crown, 1999.

Farred, Grant. "Cool as the Other Side of the Pillow: How ESPN's Sportscenter Has Changed Television Sports Talk." *Journal of Sport & Social Issues* 24, no. 2 (2000): 96–117.

Feng, Peter X. "Recuperating Suzie Wong: A Fan's Nancy Kwan-Dary." In *Countervisions: Asian-American Film Criticism,* edited by Darrel Y. Hamamoto and Sandra Liu, 40–56. Philadelphia: Temple University Press, 2000.

Foote, Stephanie. "Making Sport of Tonya: Class Performance and Social Punishment." *Journal of Sport and Social Issues* 27, no. 1 (2003): 3–17.

Foucault, Michel. *The History of Sexuality: An Introduction.* New York: Vintage, 1980.

Frankenberg, Ruth, ed. *Displacing Whiteness: Essays in Social and Cultural Criticism.* Durham, NC: Duke University Press, 1997.

Free, Marcus, and John Hughson. "Settling Accounts with Hooligans: Gender Blindness in Football Subculture Research." *Journal of Sport & Social Issues* 27, no. 2 (2003): 136–55.

Gamson, Josh. *Freaks Talk Back: Tabloid Talk Shows and Sexual Non-Conformity.* Chicago University of Chicago Press, 1998.

Gantz, Walter, and Lawrence A. Wenner. "Fanship and the Television Sports Viewing Experience." *Sociology of Sport Journal* 12 (1995): 56–74.

Ghosh, Chris. "A Guy Thing: Radio Sports Talk Shows." *Forbes,* February 22, 1999.

Giulianotti, Richard. *Football: A Sociology of the Global Game.* London: Polity Press, 1999.

Goffman, Ervin. *Forms of Talk.* Philadelphia: University of Pennsylvania Press, 1981.

Goldberg, David Theo. "Call and Response: Sports, Talk Radio, and the Death of Democracy." *Journal of Sport & Social Issues* 22, no. 2 (1998): 212–23.

Gopinath, Gayatri. "Nostalgia, Desire, Diaspora: South Asian Sexualities in Motion." *Positions* 5, no. 2 (1997): 467–89.

Gramsci, Antonio. *Selections from Prison Notebooks.* London: New Left Books, 1971.

Griffin, Pat. *Strong Women, Deep Closets: Lesbians and Homophobia in Sport.* Champaign, IL: Human Kinetics, 1998.

Grindstaff, Laura. "Abortion and the Popular Press: Mapping Media Discourse from Roe to Webster." In *Abortion Politics in the United States and Canada: Studies in Public Opinion,* edited by Ted G. Jelen and Marthe A. Chandler. Westport, CT: Praeger, 1994.

———. *The Money Shot: Trash, Class, and the Making of TV Talk Shows.* Chicago: University of Chicago Press, 2002.

Haag, Pamela. "The 50,000 Watt Sports Bar: Talk Radio and the Ethic of the Fan." *The South Atlantic Quarterly* 95, no. 2 (1996): 453–70.

Habermas, Jurgen. *The Structural Transformation of the Public Sphere,* translated by Ted Burger. Cambridge: Polity, 1989.

Halberstam, Judith. *Female Masculinity.* Durham, NC: Duke University Press, 1998.

Hall, Stuart, and Tony Jefferson. *Resistance through Rituals: Youth Subcultures in Post-War Britain.* London: Routledge, 1983.

Harris, Fran. *Summer Madness: Inside the Wacky, Wonderful World of the WNBA.* New York: Author's Choice Press, 2001.

Herbst, Susan. "On Electronic Public Space: Talk Shows in Theoretical Perspective." *Political Communication* 12 (1995): 263–74.

Hilmes, Michele. "Rethinking Radio." In *Radio Reader: Essays in the Cultural History of Radio,* edited by Michele Hilmes and Jason Loviglio, 1–20. New York: Routledge, 2002.

Hodgson, Ed. "King of Smack." *Fastbreak—The Magazine of the Phoenix Suns,* August 18, 1999, 1–5.

hooks, bell. *Feminist Theory: From Margin to Center.* Boston: South End Press, 1984.

Hughson, John. "Among the Thugs:The New Ethnographies of Football Supporting Subcultures." *International Review for the Sociology of Sport* 33, no. 1 (1998): 43–57.

Humm, Maggie. *The Dictionary of Feminist Theory,* 2nd ed. New York: Prentice-Hall, 1995.

Hutchby, Ian. *Confrontation Talk: Arguments, Asymmetries, and Power on Talk Radio.* Mahwah, NJ: Lawrence Erlbaum Associates, 1996.

Hymes, Dell H. "Models of Interaction of Language and Social Life." In *Directions in Sociolinguistics: The Ethnography of Communication,* edited by John J. Gumprez and Dell H. Hymes, 58–71. New York Holt, Rinehart, and Winston, 1972.

Jackson, Peter, Nick Stevenson, and Kate Brooks. *Making Sense of Men's Magazines.* London: Polity 2001.

Jackson, Steven. L., and David L. Andrews. "Michael Jordan and the Popular Imaginary of Post-Colonial New Zealand." In *North American Society for the Sociology of Sport,* Sacramento, CA, 1995.

Jansen, Sue C., and Donald F. Sabo. "The Sport/War Metaphor: Hegemonic Masculinity, the Persian Gulf War and the New World Order." *Sociology of Sport Journal* 11 (1994): 1–17.

Jeffords, Susan. *Hard Bodies: Masculinity in the Reagan Era.* New Brunswick, NJ: Rutgers University Press, 1993.

Jenkins, Henry. *Textual Poachers: Television Fans and Participatory Culture.* New York: Routledge, 1992.

Jhally, Sut. "Cultural Studies and the Sports/Media Complex." In *Media, Sports, and Society,* edited by Lawrence A. Wenner, 70–93. Newbury Park, CA: Sage, 1989.

Jones, Rosie. "First, a Word About Me and My Sponsor." *The New York Times*, March 21, 2004, CI.

Kearns, Jeff. "Embedded with the Kings." *The News & Review*, May 1, 2003, 1–3.

Kellner, Douglas. *Media Culture: Cultural Studies, Identity and Politics between the Modern and Postmodern.* New York: Routledge, 1995.

Kimmel, Michael. *Manhood in America: A Cultural History.* New York: Free Press, 1996.

———. "Masculinity as Homophobia." In *Theorizing Masculinities,* edited by Harry Brod and Michael Kaufman, 119–41. Thousand Oaks, CA: Sage, 1994.

King, C. Richard. "This Is Not an Indian: Situating Claims About Indianness in Sporting Worlds." *Journal of Sport & Social Issues* 28, no. 1 (2004): 3–10.

Kissling, Elizabeth. "Skating on Thin Ice: Why Tonya Harding Could Never Be America's Ice Princess." *Undercurrent: A Journal for the Analysis of the Present* 3 (1998), 1–15.

Kitzinger, Celia. *The Social Construction of Lesbianism.* Newbury Park, CA: Sage, 1987.

Klein, Alan M. "Dueling Machos: Masculinity and Sport in Mexican Baseball." In *Masculinities, Gender Relations, and Sport,* edited by Jim A. McKay, Michael A. Messner, and Donald F. Sabo, 67–86. Thousand Oaks, CA: Sage, 2000.

Lefkowitz, Daniel. "On the Mediation of Class, Race & Gender: Intonation on Sports Radio Talk Shows." *University of Pennsylvania Working Papers in Linguistics* 3, no. 1 (1996): 207–221.

Levine, Martin P., and Michael A Kimmel. *Gay Macho: The Life and Death of the Homosexual Clone.* New York: New York University Press, 1997.

Lyman, Peter. "The Fraternal Bond as a Joking Relationship." In *Men's Lives,* edited by Michael Kimmel and Michael A. Messner, 143–154. Needham Heights, MA: Allyn & Bacon, 2001.

Mariscal, Jorge. "Chicanos and Latinos in the Jungle of Sports Talk Radio." *Journal of Sport & Social Issues* 23, no. 1 (1999): 111–117.

Markovits, Andrei S., and Steven L. Hellerman. *Offside: Soccer and American Exceptionalism.* Princeton: Princeton University Press, 2001.

McChesney, Robert. *Rich Media, Poor Democracy: Communication Politics in Dubious Times.* Champaign, IL: University of Illinois Press, 1999.

McCracken, Allison. "Scary Women and Scarred Men: Suspense, Gender Trouble, and Postwar Change." In *Radio Reader: Essays in the Cultural History of Radio,* edited by Michele Hilmes and Jason Loviglio, 183–208. New York: Routledge, 2002.

McKay, Jim, Michael A. Messner, and Donald F. Sabo, eds. *Masculinities, Gender Relations, and Sport.* Thousand Oaks, CA: Sage, 2000.

McKee, Alan. *Textual Analysis: A Beginner's Guide.* Thousand Oaks, CA: Sage, 2003.

McKenzie, Robert. "Audience Involvement in the Epideictic Discourse of TV Talk Shows. " *Communication Quarterly* 48, no 2 (2000): 190–203.

McRobbie, Angela. *Feminism and Youth Culture: From Jackie to Seventeen.* Boston: Unwin, 1991.

McWhorter, Ladelle. *Bodies and Pleasures: Foucault and the Politics of Sexual Normalization.* Bloomington: Indiana University Press, 1999.

Mechling, Jay. *On My Honor: Boy Scouts and the Making of American Youth.* Chicago: University of Chicago Press, 2001.

———. "Playing Indian and the Search for Authenticity in Modern White America." *Prospects* 5 (1980): 7–33.

Mercer, Kobena. *Welcome to the Jungle: New Positions in Black Cultural Studies.* New York: Routledge, 1994.

Messner, Michael A. *Politics of Masculinities: Men in Movements.* Thousand Oaks, CA: Sage, 1997.

———. *Power at Play: Sports and the Problem of Masculinity.* Boston: Beacon Press, 1992.

———. *Taking the Field: Women, Men and Sports.* Minneapolis: University of Minnesota Press, 2002.

Messner, Michael A., Michele Dunbar, and Darnell Hunt. "The Televised Sports Manhood Formula." *Journal of Sport & Social Issues* 24, no. 4 (2000): 380–94.

Meyer, John C. "Humor in Member Narratives: Uniting and Dividing at Work." *Western Journal of Communication* 61 (1997): 188–208.

Miller, Toby. "Out at the Game." *Advocate*, June 2001a.

———. *Sportsex.* Philadelphia: University of Temple Press, 2001b.

Moores, Shaun. *Interpreting Audiences: The Ethnography of Audience Consumption.* Thousand Oaks, CA: Sage, 1993.

Morley, David. *Television Studies and Cultural Studies.* New York: Routledge, 1992.

Moss, Pamela, and C. S. Higgins. *Sounds Real: Radio in Everyday Life.* St. Lucia: University of Queensland Press, 1982.

Munoz, Jose Esteban. *Disidentifications: Queers of Color and the Performance of Politics.* Minneapolis: University of Minnesota Press, 1999.

Nakayama, Thomas, and Robert L. Krizek. "Whiteness: A Strategic Rhetoric." *Quarterly Journal of Speech*, no. 81 (1995): 291–309.

Nelson, Mariah Burton. *The Stronger Women Get, the More Men Love Football.* New York: Avon, 1994.

Newton, Judith. *From Panthers to Promise Keepers: Rethinking the Men's Movement.* Lanham, MD: Rowman & Littlefield, 2005.

Nightengale, Bob. "Reasons for Bonds' Bad Image Split between Steroids and Racism." *USA Today* (2006), http://www.usatoday.com/sports/baseball/nl/giants/2006-03-29-bonds-cover_x.html.

Nixon, Sean. "Exhibiting Masculinity." In *Representation: Cultural Representations and Signifying Practices*, edited by Stuart Hall, 291–336. Thousand Oaks, CA: Sage, 1997.

Nylund, David. *Treating Huckleberry Finn: A New Narrative Approach with Kids Diagnosed ADD/ADHD*. San Francisco: Jossey-Bass, 2000.

Oldenburg, Ray. *The Great Good Place: Cafes, Coffee Shops, Community Centers, Beauty Parlors, General Stores, Hangouts, and How They Get You through the Day*. New York: Paragon House, 1989.

O'Sullivan, Sara. "The Ryanline is Now Open: Talk Radio and the Public Sphere." In *Media Audiences in Ireland: Power and Cultural Identity*, edited by Mary J. Kelly and Barbara O'Connor, 167–190. Dublin: UCD Press, 1997.

Page, Benjamin I., and Jason Tannenbaum. "Populistic Deliberation and Talk Radio." *Journal of Communication* 46, no. 2 (1996): 33–53.

Philipsen, Gerry. "Places for Speaking in Teamstersville." *Quarterly Journal of Speech* 62 (1976): 15–25.

Pronger, Brian. *Arena of Masculinity: Sports, Homosexuality, and the Meaning of Sex*. New York: St. Martin's Press, 1990.

Radway, Janice A. *Reading the Romance: Women, Patriarchy, and Popular Literature*. Durham, NC: University of North Carolina Press, 1984.

Roediger, David. "White Looks: Hairy Apes, True Stories and Limbaugh's Laughs." *Minnesota Review* 47 (1996): 41–52.

Sabo, Donald F., and Sue C. Jansen. "Prometheus Unbound: Constructions of Masculinity in the Sports Media." In *Mediasport*, edited by Lawrence A. Wenner, 202–17. New York: Routledge, 1998.

Schiller, Herbert. *Culture, Inc.* New York: Oxford University Press, 1989.

Schwable, Michael. *Unlocking the Iron Cage: The Men's Movement, Gender, Politics, and American Culture*. New York: Oxford University Press, 1996.

Sedgwick, Eve Kosofsky. *Between Men*. New York: Columbia University Press, 1985.

———. *Epistemology of the Closet*. Berkeley: University of California Press, 1990.

Shields, David. *Black Planet: Facing Race During an NBA Season*. New York: Crown, 1999.

Simons, Herbert D. "Race and Penalized Sports Behaviors." *International Review for the Sociology of Sport* 38, no. 1 (2003): 5–22.

Smith, Maureen. "The Jim Rome Show and Negotiations of Manhood: Surviving in 'the Jungle.'" In *North American Society for the Sociology of Sport*, Indianapolis, 2002.

Snyder, John. "Handicapping the Ratings: Inside the Numbers of America's Sports Stations" (2002), www.arbitron.com/downloads/sportsradio.pdf.

Solomon, Nick. "Announcing the P.U.-Litzer Prizes for 2003." *Alternet.org* (2003), http://www.alternet.org/story.htmlSpivak, L.

Strong, Pauline Turner. "The Mascot Slot: Cultural Citizenship, Political Correctness, and Pseudo-Indian Sports Symbols." *Journal of Sport & Social Issues* 28, no. 1 (2004): 79–87.

Swan, Gary. "Roaring Success: Sponsors Fuel Nascar's Huge Popularity." *San Francisco Chronicle*, June 24, 1998, http://www.sfgate.com/cgi-bin /article.cgi?f=/chronicle/archive/1998/06/24/SP17674.DTL.

Tannen, Deborah. *You Just Don't Understand: Women and Men in Conversation*. New York: Morrow, 1990.

Tremblay, Sheryl, and William Tremblay. "Mediated Masculinity at the Millennium: The Jim Rome Show as a Male Bonding Speech Community." *Journal of Radio Studies* 8, no. 2 (2001): 271–91.

Trujillo, Nick. "Interpreting (the Work and the Talk of) Baseball: Perspectives on Ballpark Culture." *Western Journal of Communication*, no. 56 (1992): 350–71.

———. *The Meaning of Nolan Ryan*. College Station, TX: Texas A & M University Press, 1994.

———. "Hegemonic Masculinity on the Mound: Media Representations of Nolan Ryan and the American Sports Culture." In *Rhetorical Criticism: Exploration and Practices* edited by Sonja K. Foss, 181–203. Prospect Heights, IL: Waveland Press, 1994.

Vande Berg, Leah R., and Sarah Projansky. "Hoop Games: A Narrative Analysis of Televised Coverage of Women's and Men's Professional Basketball." In *Case Studies in Sport Communication*, edited by Robert S. Brown and Daniel J. O'Rourke, 27–50. Westport, CT: Praeger, 2003.

Wada, Fainaru-Wada, and Lance Willams. *Game of Shadows: Barry Bonds, Balco, and the Steroids Scandal That Rocked Professional Sports*. New York: Gotham, 2006.

Watson, Cary. "Peanuts, Popcorn and Anti-Capitalism." *Z Magazine*, April 2002, 10–12.

Wellman, Barry. "Men in Networks: Private Communities, Domestic Friendships." In *Men's Friendships*, edited by Peter Nardi, 74–114. Newbury Park, CA: Sage Publications, 1992.

Wenner, Lawrence A., ed. *Mediasport*. New York: Routledge, 1998a.

———. "The Sports Bar: Masculinity, Alcohol, Sports and the Mediation of Public Space." In *Sports and Postmodern Times: Gender, Sexuality, the Body and Sport*, edited by Genevieve Rail, 301–333. Albany: State University of New York Press, 1998b.

Whannel, Garry. *Media Sports Stars: Masculinities and Moralities*. New York: Routledge, 2002.

Wheatley, Elizabeth. "Subcultural Subversions: Comparing Discourses on Sexuality in Men's and Women's Rugby Songs." In *Women, Sport, and Culture*, edited by Susan Birrell and C. L . Cole, 286–300. Champaign, IL: Human Kinetics, 1994.

Wheaton, Belinda. "New Lads"?: Masculinities and the "New Sport" Participant." *Men and Masculinities* 2, no. 4 (2000): 434–56.

White, Michael, and David Epston. *Narrative Means to Therapeutic Ends.* New York: Norton, 1990.

Wittig, Monique. "One Is Not Born a Woman." In *The Lesbian and Gay Studies Reader,* edited by Henry Abelove, Michele Barale, and David M. Halperin, 103–09. New York: Routledge, 1993.

Wray, Matt, and Annalee Newitz. *White Trash: Race and Class in America.* New York: Routledge, 1997.

Zirin, Dave. "Burying Barry: Bonds and the Chain of Command." *ZNet* (2006), http://www.zmag.org/content/showarticle.cfm?SectionID=30& ItemID=9902.

INDEX